LINEAR FUNCTIONAL ANALYSIS

INTRODUCTION TO LEBESGUE INTEGRATION AND INFINITE-DIMENSIONAL PROBLEMS

BERNARD EPSTEIN

Professor of Mathematics
University of New Mexico

W. B. SAUNDERS COMPANY

PHILADELPHIA • LONDON • TORONTO

W. B. Saunders Company: West Washington Square
Philadelphia, Pa. 19105

12 Dyott Street
London WC1A 1DB

1835 Yonge Street
Toronto 7, Ontario

Linear Functional Analysis: An Introduction to Lebesgue Integration and Infinite Dimensional Problems SBN 0-7216-3395-1

 Press of W. B. Saunders Company. Library of Congress Catalog card number 77-92130.

Print No.: 9 8 7 6 5 4 3 2

PREFACE

In recent decades the expression "functional analysis" has come to cover a rapidly expanding and increasingly diffuse area of mathematics. One may adopt, as a rough working definition, the following: Functional analysis is that discipline which has arisen out of the attempt to extract from diverse problems of analysis their common underlying features and to develop an abstract theory which is applicable to these diverse problems as particular cases.

It has often happened in the history of mathematics that an abstract theory which has "spun off" from more concrete subject-matter is pursued for its own sake and that its practitioners forget (or never learn) the "practical" sources of the subject in which they are so deeply interested. (To cite only one example, it would take very little effort to discover, in many graduate departments of mathematics, a substantial number of students [and also some faculty members] who can quote and prove with great precision the Fredholm Alternative without realizing that it is simply an abstract version of the fundamental result in the theory of integral equations.) This is doubly unfortunate; on the one hand, a student who takes a highly abstract course may fail to realize how he might effectively apply the theory to a specific problem on which he is working, and, on the other hand, the research mathematician who works on the abstract theory without knowing its origins may be depriving himself of a rich source of interesting and important topics of investigation.

This short book is written with the purpose of introducing the reader to a very limited, and comparatively elementary, portion of the vast field of functional analysis in such a manner that he will see that the development of the subject-matter was stimulated by significant problems of concrete, or "hard," analysis, and that, conversely, the abstract theory may often be effectively employed in the study of specific concrete problems. It is my firm belief that such an introduction to functional analysis should precede a more ambitious and more abstract course in the subject, which is now required as part of the program of studies leading to the doctorate in many (probably most) departments of mathematics. Furthermore, it is my hope that this book may prove helpful to students and practitioners of disciplines other than mathematics who must make effective use of a somewhat limited mathematical training. It has become commonplace to say that when mathematicians and users of mathematics separate themselves from each other the result is that both suffer, but the fact that this remark is commonplace detracts neither from its correctness nor its pertinence. I shall be pleased if this book helps to some extent in encouraging mutually advantageous contact between the makers and the users of mathematics.

It remains to express appreciation to those persons who have helped in various ways to bring this book into being: A number of students who urged me to expand a set of lecture notes into book form; my wife, Florence, and son, David, who encouraged me and who converted scrawled sheets into typed manuscript; Dr. Fred Greenleaf of New York University, who enthusiastically and unselfishly devoted much time and energy to reading the manuscript, suggesting additional topics, detecting errors, and making many valuable suggestions for improving the exposition; Mr. Peter Renz of Reed College, who also read the manuscript and made a number of valuable suggestions; and Mr. George Fleming, the mathematics editor of W. B. Saunders Co., who, along with his capable assistants, saw the book through the many stages from handwritten notes to finished product.

Albuquerque, New Mexico BERNARD EPSTEIN

NOTATION

§6–3 denotes the third section of Chapter 6.

Theorem 6–3 denotes the third theorem of the sixth section of the current chapter; when reference is made to a theorem in another chapter, we write, for example, "Theorem 6–3 of Chapter 4." Similar notation is used for definitions, exercises, and equations.

Any item indicated by a capital letter appears in the corresponding appendix; for example, equation (H-2) is the second equation of Appendix H.

CONTENTS

CHAPTER I

METRIC SPACES

§1. SET-THEORETIC NOTATION

We shall begin by dispensing with the customary survey of intuitive or naïve set theory, for at this stage in his mathematical studies the reader has, almost certainly, already been exposed to an excessive number of such presentations. Instead, we shall content ourselves with a statement of most of the very small amount of set-theoretic notation which will be used. A few additional notations will be introduced as the need arises.

Membership and non-membership in a set will be denoted by the symbols $\in$ and $\notin$, respectively. $\bigcup_{\alpha \in I} E_\alpha$ will denote the union of the sets E_α, where α ranges over the index-set I; if the index-set is clearly understood from the context, the less explicit notation $\bigcup_\alpha E_\alpha$, or even $\bigcup E_\alpha$, will be used. If the index-set consists of the finite set of integers

$$\{1, 2, 3, \ldots, n\}$$

or the set of all positive integers $\{1, 2, 3, \ldots\}$ we shall use such self-explanatory notation as $\bigcup_{k=1}^{n} E_k$, $E_1 \cup E_2 \cup E_3 \cup \cdots \cup E_n$, $\bigcup_{k=1}^{\infty} E_k$, or $E_1 \cup E_2 \cup E_3 \cup \cdots$. For intersections we shall use analogous notations, with $\bigcup$ and $\cup$ replaced by $\bigcap$ and $\cap$, respectively. The symbol $\subset$ denotes *proper* set inclusion, while $\subseteq$ denotes *either* proper or improper set inclusion. The empty set will be denoted $\emptyset$. The difference between the sets A and B will be denoted $A - B$. (It is not required that $B \subseteq A$.) The complement of A will be denoted A^c; of course, A^c is well defined only when a "universal set" U is prescribed, either explicitly or implicitly.

§2. METRIC SPACES

Over a period of many decades, but particularly since the early part of the present century, it has been realized that many seemingly different theorems (indeed, entire theories) manifest, upon close examination, a deep resemblance, both in content and in the general pattern of development. Therefore, some of the major efforts of mathematical investigators have been directed toward discovering the underlying reasons for the fundamental resemblances which may be concealed by superficial differences. One of the most basic and useful ideas that has emerged from these efforts is that of a metric space. After presenting the definition, we shall give a number of illustrative examples of varying degrees of difficulty and significance. Then in the remaining sections of this chapter and in later portions of this book, we shall illustrate the use of this concept and of various related concepts in the study of numerous highly significant problems of analysis.

Definition 1: *A metric space is a non-empty set M in which there is defined, for each ordered pair (f, g) of elements of M, a real number $\rho(f, g)$, called the distance from f to g, satisfying the following conditions:*
(i) $\rho(f, g) = \rho(g, f)$ *(symmetry)*;
(ii) $\rho(f, g) > 0$ *if* $f \neq g$, $\rho(f, g) = 0$ *if* $f = g$;
(iii) $\rho(f, h) \leqslant \rho(f, g) + \rho(g, h)$ *(triangle inequality). (Strictly speaking, a metric space is a pair of objects: a set M and a real-valued function ρ defined on** $M \times M$. *Therefore, some notation such as (M, ρ) or M_ρ should be employed; however, if [as is usually the case] a single distance-function, or metric, ρ is under consideration, one may denote the metric space by the single letter M.)*

Examples

(*a*) Trivially, the set consisting of a single point f, with $\rho(f, f) = 0$, is a metric space.

(*b*) Let M be any non-empty set, and let $\rho(f, g) = 0$ if $f = g$, while $\rho(f, g) = 1$ if $f \neq g$.

(*c*) The real number system, with the distance $\rho(f, g)$ defined as $|f - g|$. This is certainly the most important single example of a metric space. It will be denoted by the letter R.

(*d*) The rational number system, with distance defined as in R. We denote this space by the letter Q (to suggest the word "quotient," since the rational numbers are quotients of integers).

(*e*) The complex number system, with the distance $\rho(f, g)$ defined as $|f - g|$. We denote this metric space by the letter $\mathfrak{C}$.

---

* The symbol $\times$ denotes the cartesian product: $A \times B$ is the collection of all *ordered* pairs (a, b), where $a \in A$ and $b \in B$.

(*f*) The class of all continuous real-valued functions defined on the bounded closed interval $[a, b]$, with $\rho(f, g) = \max_{a \leqslant x \leqslant b} |f(x) - g(x)|$. This space is denoted $C([a, b])$. (The letter C suggests "continuous.")

(*g*) The class of all bounded, but not necessarily continuous, real-valued functions defined on the interval $[a, b]$, with*

$$\rho(f, g) = \sup_{a \leqslant x \leqslant b} |f(x) - g(x)|.$$

(*h*) The same class of functions appearing in (*f*), but with $\rho(f, g) = \int_a^b |f(x) - g(x)|\, dx$.

(*i*) The class of all bounded real-valued functions defined and Riemann-integrable on the interval $[a, b]$ does *not* become a metric space with the definition of $\rho(f, g)$ employed in (*h*), for the "distance" between the function $f \equiv 0$ and the function g, where

$$g(x) = \begin{cases} 1, & \text{when } x = a, \\ 0, & \text{otherwise,} \end{cases}$$

would be zero, although f and g are different members of the class.

(*j*) The set of all ordered triples of real numbers (a_1, a_2, a_3), where the distance between (a_1, a_2, a_3) and (b_1, b_2, b_3) is defined as

$$\max \{|a_1 - b_1|, |a_2 - b_2|, |a_3 - b_3|\}.$$

This definition extends in an obvious manner to ordered n-tuples of real numbers, where n is any fixed positive integer.

(*k*) The same set that appears in (*j*), but with the distance between (a_1, a_2, a_3) and (b_1, b_2, b_3) defined as $|a_1 - b_1| + |a_2 - b_2| + |a_3 - b_3|$.

(*l*) The same set that appears in (*j*), but with the distance between (a_1, a_2, a_3) and (b_1, b_2, b_3) defined as $[(a_1 - b_1)^2 + (a_2 - b_2)^2 + (a_3 - b_3)^2]^{1/2}$. (The fact that the triangle inequality holds is not quite obvious, but a proof is sketched in Exercises 4 and 5; an extended version of the result of Exercise 4 [Schwarz inequality] plays a vital role in much of the theory of linear spaces, as will be seen in later chapters.)

(*m*) Once again we take the same set that appears in (*j*), (*k*), and (*l*), and we choose a fixed number p exceeding or equaling one. We define the distance between (a_1, a_2, a_3) and (b_1, b_2, b_3) as $[|a_1 - b_1|^p + |a_2 - b_2|^p + |a_3 - b_3|^p]^{1/p}$. The proof of the validity of the triangle inequality is deferred to §3–6. Note that the cases $p = 1$ (which is elementary) and $p = 2$ are covered in (*k*) and (*l*), respectively.

---

* We use the term "max" only when the supremum (least upper bound) is actually attained, as it certainly is in (*f*), but not always in (*g*).

Exercises

1. Give a reasonable example of a situation in which one encounters a distance function which violates the condition (*i*) of Definition 1. (Hint: Consider a number of cities connected by a network of roads.)

2. Let M be a metric space and let N be a non-empty subset of M, with distance between elements of N defined as in M. Show that N is then also a metric space. Whenever we refer to a non-empty subset of a metric space as a metric space in its own right, it is understood that the metric is defined in this way. (That is, the metric in N is simply the restriction of the original metric to $N \times N$.)

3. Consider the metric spaces described in Examples (*f*) and (*h*); show that a pair of functions, f and g, may be close to each other in one metric and far apart in the other.

4. Let $a_1, a_2, \ldots, a_n$ and $b_1, b_2, \ldots, b_n$ be real numbers, at least one of the b's being different from zero. Show that the quadratic equation

$$\sum_{k=1}^{n} (a_k - b_k x)^2 = 0$$

cannot have two distinct real roots. Deduce from this fact the Cauchy-Schwarz inequality

$$\left\{\sum_{k=1}^{n} a_k b_k\right\}^2 \leqslant \left\{\sum_{k=1}^{n} a_k^2\right\}\left\{\sum_{k=1}^{n} b_k^2\right\}.$$

(Of course, if the b's all vanish there is nothing to prove.)

5. Use the Cauchy-Schwarz inequality to establish (for arbitrary real numbers $a_1, a_2, \ldots, a_n$ and $b_1, b_2, \ldots, b_n$) the inequality

$$\left\{\sum_{k=1}^{n} (a_k + b_k)^2\right\}^{1/2} \leqslant \left\{\sum_{k=1}^{n} a_k^2\right\}^{1/2} + \left\{\sum_{k=1}^{n} b_k^2\right\}^{1/2}.$$

From this result demonstrate, in turn, the inequality (valid for arbitrary real numbers $a_1, a_2, \ldots, a_n$, $b_1, b_2, \ldots, b_n$, $c_1, c_2, \ldots, c_n$)

$$\left\{\sum_{k=1}^{n} (a_k - c_k)^2\right\}^{1/2} \leqslant \left\{\sum_{k=1}^{n} (a_k - b_k)^2\right\}^{1/2} + \left\{\sum_{k=1}^{n} (b_k - c_k)^2\right\}^{1/2}.$$

(Note that for $n = 3$ this demonstrates the triangle inequality for the space defined in Example (*l*).)

6. For arbitrary *complex* numbers prove the inequalities

$$\left|\sum_{k=1}^{n} a_k b_k\right|^2 \leqslant \left\{\sum_{k=1}^{n} |a_k|^2\right\}\left\{\sum_{k=1}^{n} |b_k|^2\right\},$$

$$\left\{\sum_{k=1}^{n} |a_k + b_k|^2\right\}^{1/2} \leqslant \left\{\sum_{k=1}^{n} |a_k|^2\right\}^{1/2} + \left\{\sum_{k=1}^{n} |b_k|^2\right\}^{1/2},$$

$$\left\{\sum_{k=1}^{n} |a_k - c_k|^2\right\}^{1/2} \leqslant \left\{\sum_{k=1}^{n} |a_k - b_k|^2\right\}^{1/2} + \left\{\sum_{k=1}^{n} |b_k - c_k|^2\right\}^{1/2}.$$

§3. ELEMENTARY TOPOLOGY OF METRIC SPACES

In this and the next section we shall encounter certain concepts which are obvious extensions of those arising in the study of the fundamental principles of real analysis, and so our discussions will justifiably be quite concise.

Given a metric space M, a subset S of M, and an element x of M (not necessarily contained in S), we say that x is a *limit-point* of S if for every positive number ϵ there exists at least one element y belonging to S *and distinct from* x such that $\rho(x, y) < \epsilon$. Alternatively (cf. Exercise 1), x is a limit-point of S if for every positive number ϵ there exist infinitely many members of S whose distance from x is less than ϵ. (In particular if S consists of only a finite number of points, it has no limit-point.) The set S is said to be *closed* if it contains all its limit-points. (Of course, if S has no limit-points, it *contains* all its limit-points, and so it is closed; in particular, any finite subset of M is closed.)

Having defined a limit-point, we now quote without proof the following theorem, which is one of the foundation stones of real analysis. If the reader is not thoroughly acquainted with this theorem and its proof he should study it carefully in a text dealing with the fundamentals of real analysis.

Theorem I (Bolzano-Weierstrass): *Let S be an infinite bounded* subset of the real number system R. Then S has (but does not necessarily contain!) at least one limit-point.*

A point y belonging to the subset S of M is termed an *inner point* of S if there exists some positive number δ such that every point of M whose distance from y is less than δ is a member of S. The set S is said to be *open* if all its points are inner points (of S). The relation between open and closed sets is the same as that which holds in R: the set S is open iff† S^c is closed. (Cf. Exercise 2.)

* A subset S of R is said to be *bounded* if there exist real numbers a and b such that the inequalities $a \leqslant x \leqslant b$ hold for every member x of S. Equivalently, S is said to be bounded if there exists a positive number c such that the inequality $|x - y| \leqslant c$ holds for all members x and y of S.

† If and only if.

For any positive number ϵ and for any point x of M, $S_\epsilon(x)$ and $S_\epsilon[x]$ denote the subsets of M consisting of all points y satisfying the inequalities $\rho(x, y) < \epsilon$ and $\rho(x, y) \leqslant \epsilon$, respectively. Such sets are called *open balls* and *closed balls*, respectively. (The justification of these terms is the subject of Exercise 9.) The point x is termed the center of the ball, and ϵ is termed the radius.

In a very reasonable sense, open balls are the building blocks from which all open sets are constructed. This statement is made precise in the following theorem, whose simple proof is left to the reader as Exercise 10.

Theorem 2: *A subset S of a metric space is open iff it can be expressed as the union of a collection (not necessarily finite or countably infinite) of open balls; equivalently, S is open iff for every element x in S there exists a positive number ϵ such that $S_\epsilon(x) \subseteq S$.*

The *interior* of a subset S, denoted int S, of M is defined as the set of all inner points of S; int S is always an open set, and S is open iff $S = \text{int } S$. (Cf. Exercise 3.) Similarly, the *closure* of S is defined as the union of S and the set of all its limit-points; it is denoted $\bar{S}$. The set $\bar{S}$ is always closed, and S is closed iff $S = \bar{S}$; hence, for any set S, $\bar{\bar{S}} = \bar{S}$. (Cf. Exercise 4.)

At this point we draw the reader's attention to Exercise 4–10, which ties up the concepts of limit-point and closure with that of a convergent sequence.

We conclude this section with an important definition: The subset S of M is called *dense* if $\bar{S} = M$.

Exercises

1. Prove the equivalence of the two given definitions of a limit-point.
2. Prove the assertion that S is open iff S^c is closed.
3. Prove that S is open iff $S = \text{int } S$, and that, for any set S, int (int S) = int S.
4. Prove that S is closed iff $S = \bar{S}$, and that, for any set S, $\bar{\bar{S}} = \bar{S}$.
5. Prove that any metric space M (considered as a subset of itself) is both open and closed; also that the empty set $\emptyset$ (considered as a subset of M) is both open and closed.
6. Prove that $\emptyset$ and R are the only subsets of R which are both open and closed.
7. In contrast to the preceding exercise, prove that every subset of the metric space defined in Example (b) of §2 is both open and closed.

8. Prove that the union of a *finite* collection of closed subsets and the intersection of *any* collection of closed subsets are closed. Then state and prove corresponding assertions concerning open subsets.

9. Prove that open and closed balls are open and closed sets, respectively.

10. Prove Theorem 2.

11. Prove that $\overline{S_\epsilon(x)} \subseteq S_\epsilon[x]$. Show, by means of a simple example, that the inclusion may be proper.

§4. CONVERGENCE AND COMPLETENESS

The reader is certainly acquainted with the concept of a sequence of real numbers; a formal definition of such a *sequence* is that it is a function (mapping, transformation) from the set of positive integers into the set of real numbers, but informally one may think of a non-terminating succession of real numbers labeled with the positive integers. Similarly, we can define a sequence of objects which are members of any non-empty set. (It should be emphasized that repetitions are permitted; the tenth term may be the same as the fifth, and it may even happen that all terms of the sequence coincide.) We shall, of course, be interested in sequences whose terms are points of a given metric space M; more briefly, sequences in M.

Such a sequence, say $x_1, x_2, x_3, \ldots$, is said to *converge* to the point x if for every positive number ϵ there exists a positive integer N (depending usually on ϵ) such that $\rho(x, x_n) < \epsilon$ for *all* (not merely *some*) integers n exceeding N. The sequence is said to be convergent, or to converge, if there exists a point x to which the sequence converges. It is a very elementary but important fact that a sequence which converges to x cannot also converge to some other point y. For suppose that this could occur. Then for any positive number ϵ, we could choose an integer N_1 such that $\rho(x, x_n) < \epsilon$ whenever $n > N_1$, and similarly we could choose an integer N_2 such that $\rho(y, x_n) < \epsilon$ whenever $n > N_2$. Choosing an integer n exceeding *both* N_1 and N_2, we would then obtain, by the triangle inequality, $\rho(x, y) \leqslant \rho(x, x_n) + \rho(y, x_n) < \epsilon + \epsilon = 2\epsilon$, and so, because of the arbitrariness of ϵ, $\rho(x, y) = 0$, or $x = y$. Thus, we are justified in speaking of x as *the* limit, rather than *a* limit, of a convergent sequence. We often write $x_n \to x$, or $\lim_{n\to\infty} x_n = x$.

Given a sequence converging to some limit x, we can also assert that terms far out in the sequence are very close to each other; for, given any positive number ϵ, we can choose an integer N such that whenever $n > N$, $\rho(x, x_n) < \epsilon/2$. Choosing integers p and q exceeding N, we obtain, again by the triangle inequality, $\rho(x_p, x_q) \leqslant \rho(x_p, x) + \rho(x_q, x) < \epsilon/2 + \epsilon/2 = \epsilon$.

One might suppose that the preceding reasoning is reversible; i.e., that if for every positive ϵ there exists an integer N such that $\rho(x_p, x_q) < \epsilon$ whenever p and q both exceed N, then the sequence is convergent. This supposition is, however, false. To prove this, let us consider the metric space R and in this space form the convergent sequence 1.4, 1.41, 1.414, . . . consisting of successive decimal approximations to $\sqrt{2}$. If, however, we consider the given sequence as a sequence in Q rather than in R, it is *not* convergent, for there is no number in Q to which the terms of the sequence come closer and closer. (Remember that $\sqrt{2}$ is irrational, or, in other words, that it does not exist if the universe is Q rather than R.)

We are therefore led to define a sequence (in a metric space) to be a *Cauchy sequence* if for every positive number ϵ there exists an integer N such that $\rho(x_p, x_q) < \epsilon$ whenever p and q both exceed N; and then we define a metric space M to be *complete* if every Cauchy sequence in M is convergent. Referring back to the previous paragraph, we see that the sequence 1.4, 1.41, 1.414, . . . is a Cauchy sequence, but not a convergent sequence, in Q, and so the metric space Q is not complete. The reader is presumably aware that R, in contrast, *is* complete, and that this is the fundamental reason why R, rather than Q, serves as the setting for real analysis.

Exercises

1. A point x is said to be *isolated* if $\{x\}$ (i.e., the set consisting of the single point x) is open. Prove that if x is not isolated there exists a sequence, consisting of points distinct from x and from each other, which converges to x.

2. Prove that a non-empty subset of a complete metric space is complete iff it is closed.

3. Prove that every subsequence* of a Cauchy sequence is Cauchy and that every subsequence of a convergent sequence is convergent to the same limit.

4. Suppose that a Cauchy sequence possesses a convergent subsequence. Prove that the original sequence is also convergent.

5. Let M be a complete metric space and let $F_1, F_2, \ldots$ be non-empty *nested* subsets of M. (That is, $\cdots \subseteq F_3 \subseteq F_2 \subseteq F_1$.)

---

* Let $a_1, a_2, a_3, \ldots$ be any sequence and let $n_1, n_2, n_3, \ldots$ be a strictly increasing sequence of positive integers (i.e., $0 < n_1 < n_2 < n_3 < \cdots$). Then the sequence $a_{n_1}, a_{n_2}, a_{n_3}, \ldots$ is a *subsequence* of the original sequence. Thus, a subsequence is obtained by removing some, perhaps infinitely many, of the terms of the given sequence, provided that an infinite number of terms remain and the order in which the terms appear is not altered.

Show that if each of the F's is closed and if the diameter* of F_k approaches zero with increasing k, then $\bigcap_{k=1}^{\infty} F_k$ is not empty—in fact, the intersection consists of exactly one point. (This is the Cantor intersection theorem.)

6. Prove (by considering various subsets of R) that the assertion made in the preceding exercise fails if either the completeness condition, the closedness condition, or the condition on the diameters is omitted.

7. Let M be any metric space and let two sequences in M (not necessarily convergent, or even Cauchy), say $x_1, x_2, x_3, \ldots$ and $y_1, y_2, y_3, \ldots$, be called equivalent if $\lim_{n\to\infty} \rho(x_n, y_n) = 0$. First, show that this is indeed an equivalence relation.† Second, show that if one of two equivalent sequences is convergent, then the other is also convergent and the two sequences have the same limit.

8. (Continuation of Exercise 7). Let A be the class of all Cauchy sequences in M, and let s_1 and s_2 be two members of A:

$$s_1 = x_1^{(1)}, x_2^{(1)}, x_3^{(1)}, \ldots,$$

$$s_2 = x_1^{(2)}, x_2^{(2)}, x_3^{(2)}, \ldots.$$

Show that $\lim_{n\to\infty} \rho(x_n^{(1)}, x_n^{(2)})$ exists and is unaltered if s_1 and s_2 are replaced by equivalent sequences t_1 and t_2, respectively; also, show that this limit is zero iff s_1 and s_2 are equivalent.

Finally, show how these results can be used to define a complete metric space whose elements are the various equivalence classes of Cauchy sequences in M and such that the elements of M, in a reasonable sense, constitute a dense subset. (This new space is termed the *sequential completion* of M; if M is complete, then it is essentially indistinguishable from its sequential completion.)

9. Show that the diameter of a subset equals the diameter of its closure.

---

* The *diameter* of a non-empty subset S of a metric space M is defined as $\sup_{x,y\in S} \rho(x, y)$. The diameter may be infinite, of course, as in the case $M = R$, $S = Q$.

† A relation $\sim$ defined on a non-empty set A is termed an *equivalence relation* if $x \sim x$ for every x in A, if $x \sim y$ implies $y \sim x$, and if $x \sim y$ and $y \sim z$ implies $x \sim z$. (For the definition of a relation on a non-empty set, the reader may refer to the first two sentences of Appendix A.) Any equivalence relation in A determines a *partition* of A (i.e., a decomposition of A into disjoint subsets such that elements x and y belong to the same subset iff $x \sim y$). These subsets are called *equivalence classes* of A (with respect to the given equivalence relation).

10. (*a*) Let S be a subset of a metric space M. Show that the point x is a limit-point of S iff there exists a sequence in $S - \{x\}$ which converges to x.
(*b*) Show that S is closed iff every point in M which is the limit of a sequence in S belongs to S.
(*c*) Show that $\bar{S}$ is the set of all points in M which are limits of sequences in S.

11. Show that the Bolzano-Weierstrass theorem can be reformulated as follows: Given any bounded infinite subset S of R, it is possible to extract from S a convergent sequence $x_1, x_2, x_3, \ldots$ whose terms are all distinct. (Of course, the limit of this sequence need not belong to S.)

12. Prove that the diameter of either an open ball $S_\epsilon(x)$ or a closed ball $S_\epsilon[x]$ cannot exceed 2ϵ. Show by means of a simple example that the diameter may be less than 2ϵ. Can the diameter of a ball equal zero? If so, what can you say about the point x?

§5. THE CONTRACTION THEOREM

Many mathematical problems may be stated in the form $x = Tx$, where the unknown x is some kind of mathematical object (perhaps, but not necessarily, a real number), while Tx denotes the result of carrying out some operation on x. We deliberately content ourselves with this rather vague statement at this time. We shall first give a few simple illustrations and then proceed to develop a very general method which is often effective in solving such problems.

(*i*) Let us begin with the extremely trivial algebraic problem $x = \frac{1}{2}x + 5$. We may think of T as being the operation of multiplying any real number by $\frac{1}{2}$ and adding 5 to the result. If we guess that $x = 10$ is the solution (or at least *a* solution), we can confirm the correctness of this guess by replacing x on both sides of the equation by 10 and observing that we obtain $10 = 10$. On the other hand, if we guess that $x = 11$ is a solution, we shall obtain the false statement $11 = 10.5$, when we replace x on both sides by 11. However, if we replace x by 11 only on the right side, we find that the right side reduces to 10.5, and this may suggest trying 10.5 as a new (and hopefully better) guess. Indeed, the right side then becomes 10.25, and by repeatedly replacing x on the right side by the preceding numerical result, it is quite clear that we obtain a sequence of numbers which converges to 10, which is indeed a solution (in fact, the only solution) of the problem. The reader should have no difficulty in showing that any initial guess, no matter how far from the number 10, will give rise, when the preceding procedure is followed, to a sequence of "improved guesses" which converges to 10.

(*ii*) Suppose we convert the preceding problem into the entirely equivalent form $x = 2x - 10$. It is not difficult to see that if we employ any initial guess other than 10, the succeeding guesses will get worse rather than better; for example, if we choose 10.01 as the initial guess, the succeeding guesses will form the sequence 10.02, 10.04, 10.08, 10.16, and so forth. Quite clearly the factor 2 appearing before the x on the right side is responsible for doubling the error at each successive stage, while the factor $\frac{1}{2}$ appearing in the earlier formulation is responsible for halving the error at each successive stage.

(*iii*) Let us write the equation $x^2 - 3x + 2 = 0$ (which has the two solutions $x = 1, 2$) in the form $x = x^2/3 + \frac{2}{3}$. If we choose as an initial guess $x = 1.5$ and substitute into the right side, we obtain the succeeding guess $x = \frac{17}{12}$, which is smaller than the initial guess. By inserting the new value into the right side, we obtain a still smaller guess, and it is not difficult to see that by repeating this procedure indefinitely we obtain a sequence which converges to 1, which is indeed one of the two solutions of the given problem. On the other hand, if we choose as the initial guess the number 3, the resulting sequence will increase without bound. Finally, we remark that, unless the initial guess is *exactly* 2 or -2, the resulting sequence will certainly not converge to the solution 2. (Cf. Exercise 4.) Thus, in contrast to example (*i*), where the method described *always* succeeds, and in contrast to example (*ii*), where the method described *always* fails unless the initial guess is exactly correct, we have in the present situation a much more delicate state of affairs.

These three simple problems should suffice to give the reader an idea of the kind of theorem we are aiming at.

Theorem I (Contraction Theorem): *Let M be a complete metric space and let T be a function from M into M; i.e., for every x in M, Tx is a uniquely determined member of M. Suppose that T is contracting; i.e., there exists a fixed positive number k less than unity such that, for any elements x, y of M, the inequality $\rho(Tx, Ty) \leqslant k\rho(x, y)$ holds. Then the equation $x = Tx$ possesses a unique solution in M, and it can be obtained by iteration, or successive approximation, beginning with any element y of M. By this we mean that, no matter how y is chosen in M, the sequence* y, Ty, T^2y, T^3y, . . . converges to the unique solution of the preceding equation.*

PROOF: First we establish that the given problem has *at most* one solution. Suppose, on the contrary, that $x_1 = Tx_1$ and $x_2 = Tx_2$, where $x_1 \neq x_2$. Then $0 < \rho(x_1, x_2) = \rho(Tx_1, Tx_2) \leqslant k\rho(x_1, x_2)$, and so

$$(1 - k)\rho(x_1, x_2) \leqslant 0;$$

* Of course, T^2y means $T(Ty)$, T^3y means $T(T^2y)$, and so forth.

this is impossible, since the left side of the inequality is the product of two *positive* numbers. Thus the solution, if any exists, is unique.

Now, we observe that $\rho(T^2f, T^2g) = \rho(T(Tf), T(Tg)) \leqslant k\rho(Tf, Tg) \leqslant k^2\rho(f, g)$, and by an obvious induction we obtain $\rho(T^nf, T^ng) \leqslant k^n\rho(f, g)$ for any positive integer n. If $0 < m < n$, we then obtain, from the triangle inequality,

$$\begin{aligned}\rho(T^my, T^ny) &\leqslant \rho(T^my, T^{m+1}y) + \rho(T^{m+1}y, T^{m+2}y) + \cdots + \rho(T^{n-1}y, T^ny)\\ &\leqslant k^m\rho(y, Ty) + k^{m+1}\rho(y, Ty) + \cdots + k^{n-1}\rho(y, Ty)\\ &= \frac{k^m - k^n}{1-k}\rho(y, Ty) \leqslant \frac{k^m}{1-k}\rho(y, Ty).\end{aligned}$$

Given $\epsilon > 0$, we can therefore choose $N(\epsilon)$ so large that whenever $N < m < n$ the inequality $\rho(T^my, T^ny) < \epsilon$ holds. Thus, the sequence y, Ty, $T^2y, \ldots$ is Cauchy, and, since M is complete,* by hypothesis, this sequence must converge to some element x. Furthermore, for any n, $\rho(x, Tx) \leqslant \rho(x, T^ny) + \rho(T^ny, T^{n+1}y) + \rho(T^{n+1}y, Tx) \leqslant \rho(x, T^ny) + k^n\rho(y, Ty) + k\rho(x, T^ny)$. As $n \to \infty$, all three terms on the right approach zero, and since the left side of the inequality, namely $\rho(x, Tx)$, is independent of n, it follows that $\rho(x, Tx) = 0$, or $x = Tx$. The proof is thus complete.

For obvious reasons, the element x whose existence has just been demonstrated is called a fixed point of the function, or operator, T. Much of modern analysis (particularly non-linear analysis) is devoted to the demonstration of the existence of fixed points of functions defined on various kinds of spaces (often, but not always, metric spaces). The preceding theorem, simple though it may appear, serves as a prototype of much more difficult theorems; furthermore, as we shall illustrate in the next section, it is itself a tool of great usefulness and power.

Exercises

1. Show, by means of a simple example, that it does *not* suffice, in the contraction theorem, to replace the hypothesis $k < 1$ by the weaker hypothesis $k \leqslant 1$, nor does it suffice to assume that $\rho(Tx, Ty) < \rho(x, y)$ whenever $x \neq y$. (That is to say, the ratio $\rho(Tx, Ty)/\rho(x, y)$ must be bounded away from unity as x and y range over M, nor merely less than unity.)

2. Show that some power of T may be contracting even though T itself is not contracting.

---

* It may appear that the hypothesis of completeness is employed simply in order to enable us to push through the proof. However, Exercise 6 shows that this hypothesis is indispensable.

3. (Continuation of Exercise 2). Suppose that T^p is contracting for some integer $p > 1$; as before, we assume that M is complete. Show that the sequence $y, Ty, T^2y, \ldots$ converges, regardless of the choice of y, and that the limit of the sequence, x, is a fixed point of T (not merely of T^p); furthermore, show that T has only one fixed point, even though T is *not* assumed to be contracting.

4. Prove the assertion made in example (*iii*) to the effect that, except for two particular initial guesses, the solution $x = 2$ will never be obtained by the iterative procedure.

5. Let M be the subset $[0, 1.4]$ of R and let $Tx = x^2/3 + \frac{2}{3}$ for every x in M. Show that all hypotheses of the contraction theorem are satisfied, so that the equation $x^2 - 3x + 2 = 0$ has exactly one solution in M.

6. Let M be the metric space consisting of all rational numbers in the closed interval $[\frac{1}{2}, \frac{3}{2}]$ (with the metric derived from R), and let $Tx = x + (2 - x^2)/10$. Show that all the hypotheses of the contraction theorem except the completeness of M are satisfied and that T has no fixed point.

§6. APPLICATION OF THE CONTRACTION THEOREM TO DIFFERENTIAL EQUATIONS

In this section we give one application of major significance of the contraction theorem. The basic problem of the entire theory of ordinary differential equations is the following: Given a function $f(x, y)$ defined in some domain D (open, connected, non-empty set) of the plane and given an interior point (x_0, y_0) of D, does there exist a solution of the differential equation $dy/dx = f(x, y)$ passing through this point—that is, a differentiable function $g(x)$ defined in some interval $I = [c, d]$ containing x_0 in its interior such that $g(x_0) = y_0$ and $dg/dx = f(x, g(x))$ for all x in the interval I? Furthermore, if such a solution exists, is it unique? Simple examples (such as $f(x, y) = 0$, $f(x, y) = 1$, $f(x, y) = 1 + y^2$), where the problem can be solved by elementary techniques, suggest that the answer to both questions is affirmative. On the other hand, it is obvious that some reasonable hypotheses must be imposed in order to guarantee the existence of a solution; for example, if $f(x, y)$ equals 0 whenever x is rational and equals 1 whenever x is irrational, it is clear that no solution exists in any interval, however small. This simple example suggests that some hypothesis involving continuity of $f(x, y)$ should be imposed. On the other hand, the case $f(x, y) = 2\,|y|^{1/2}$ shows that mere continuity is insufficient to guarantee uniqueness. (Cf. Exercise 1.)

The following theorem, while not the strongest of its type that is known, plays a vital role in the theory of ordinary differential equations. While it should be part of the analytical arsenal of every mathematician, we present it here as an excellent illustration of how the contraction theorem can often be used to establish, on the one hand, both existence and uniqueness of a solution to a given problem and to provide, on the other hand, a constructive procedure for obtaining this solution.

Theorem 1 (Cauchy-Picard): *Let $f(x, y)$ be real-valued and continuous in some domain D containing the point (x_0, y_0), and suppose that $|f(x, y_2) - f(x, y_1)| \leqslant K\,|y_2 - y_1|$, where K is a positive constant,* whenever (x, y_1) and (x, y_2) belong to D. Then there exists some interval $I = [x_0 - a, x_0 + a]$ and a function $g(x)$ defined and continuously differentiable in this interval and satisfying the conditions $g(x_0) = y_0$ and $dg/dx = f(x, g(x))$. Furthermore, there is only one function satisfying all these conditions.*

PROOF: If the function g satisfies the given differential equation and the condition $g(x_0) = y_0$, we find by integration that g must satisfy the *integral* relation

$$g(x) = y_0 + \int_{x_0}^{x} f(\xi, g(\xi))\, d\xi;$$

conversely, if g satisfies this relation it must satisfy both the given differential equation and the condition $g(x_0) = y_0$. Thus, the preceding integral relation is entirely equivalent to the given differential equation *together with* the prescribed initial condition. This fact suggests very strongly that the function g should be thought of as a fixed point (in a suitable space of functions) of a certain integral operator. We now proceed to follow up this idea in detail.

First, choose numbers α, β which are positive but so small that the closed set S defined by the inequalities $|x - x_0| \leqslant \alpha$, $|y - y_0| \leqslant \beta$ is contained in D, and let $m = \max_{(x,y)\in S} |f(x, y)|$. Then let

$$a = \min\,\{\alpha, \beta/m, 2/(3K)\}$$

---

* A function $f(x, y)$ satisfying such a condition is said to satisfy a *Lipschitz condition*, or to be Lipschitz continuous, and any number K which can be employed in the preceding inequality is termed a Lipschitz constant of $f(x, y)$.

and let S_1 be the closed subset of D defined by the inequalities $|x - x_0| \leqslant a$, $|y - y_0| \leqslant \beta$. (Clearly, $S_1 \subseteq S$, and $|f(x, y)| \leqslant m$ everywhere in S_1.) Now let M be the class of all real-valued continuous functions defined on $[x_0 - a, x_0 + a]$ whose range* is in the interval $[y_0 - \beta, y_0 + \beta]$, and let the distance between any two members of M be defined by the equation $\rho(f, g) = \max_{|x-x_0|\leqslant a} |f(x) - g(x)|$. M is easily shown to be a *complete* metric space. (Cf. Exercise 2.)

Next, we define Tg, for every $g \in M$, as follows: $Tg = h$, where $h(x) = y_0 + \int_{x_0}^{x} f(\xi, g(\xi))\, d\xi$, $x \in [x_0 - a, x_0 + a]$. We note that this definition is meaningful, for $f(\xi, g(\xi))$ is continuous; hence the integral is well defined and depends continuously on x, so that $h(x)$ is continuous on $[x_0 - a, x_0 + a]$. Furthermore, if g_1 and g_2 are any two members of M, we have

$$\left\{y_0 + \int_{x_0}^{x} f(\xi, g_1(\xi))\, d\xi\right\} - \left\{y_0 + \int_{x_0}^{x} f(\xi, g_2(\xi))\, d\xi\right\}$$
$$= \int_{x_0}^{x} \{f(\xi, g_1(\xi)) - f(\xi, g_2(\xi))\}\, d\xi,$$

and so

$$\begin{aligned}\rho(Tg_1, Tg_2) &\leqslant \max_{|x-x_0|\leqslant a} \left|\int_{x_0}^{x} \{f(\xi, g_1(\xi)) - f(\xi, g_2(\xi))\}\, d\xi\right| \\ &\leqslant \max_{|x-x_0|\leqslant a} \left|\int_{x_0}^{x} K\, |g_1(\xi) - g_2(\xi)|\, d\xi\right| \\ &\leqslant Ka \cdot \rho(g_1, g_2) \leqslant \tfrac{2}{3}\rho(g_1, g_2).\end{aligned}$$

Finally, we observe that

$$|h(x) - y_0| \leqslant \left|\int_{x_0}^{x} |f(\xi, g(\xi))|\, d\xi\right| \leqslant ma \leqslant \beta,$$

so that Tg belongs to M whenever g does; in other words, T maps M into M. Thus, we are assured by the contraction theorem that there exists one and only one member g of M satisfying the equation $g = Tg$, which is identical with the preceding integral relation.

Finally, it is evident that any function $g(x)$ satisfying in the interval $[x_0 - a, x_0 + a]$ the differential equation $dy/dx = f(x, y)$ and the condition $g(x_0) = y_0$ *must* be a member of M and must satisfy $g = Tg$. Hence the proof of the entire theorem is complete.

---

* We remind the reader that the *range* of a function is the set of all values of the function; that is, the object y belongs to the range of the function f iff there exists an object x such that $f(x)$ is defined and equals y.

Exercises

1. Show that the differential equation $dy/dx = 2|y|^{1/2}$ is satisfied by the functions $g_1(x) = 0$, $g_2(x) = x|x|$, $g_3(x) = \max\{0, x|x|\}$, and $g_4(x) = \min\{0, x|x|\}$ and that all four of these functions satisfy the initial condition $g_k(0) = 0$.

2. Prove that $C([a, b])$, as defined in §2, is a *complete* metric space; furthermore, prove that the subset of $C([a, b])$ consisting of functions satisfying the inequalities $c \leqslant f(x) \leqslant d$ for all $x \in [a, b]$ is also a complete metric space. Is this still true if we replace $c \leqslant f(x) \leqslant d$ by $c < f(x) < d$?

§7. THE BAIRE CATEGORY THEOREM

At the end of §3 we defined a subset S of a metric space M to be dense if $\bar{S} = M$. It should be immediately evident (cf. Exercise 1) that the following definition is entirely equivalent: The subset S is dense if for every *non-empty* open subset G of M the intersection $S \cap G$ is non-empty. In other words, a dense subset of M penetrates every non-empty open subset of M. It may be worth pointing out that it is possible for both S and its complement to be dense; for example, both Q and $R - Q$ are dense subsets of R.

We now define a subset T of a metric space M to be *nowhere dense* iff $\bar{T}$ contains no open ball; this, in turn, is clearly equivalent to the assertion that the interior of $\bar{T}$ is empty. We state the three following theorems, leaving the proofs to the reader as Exercise 2.

Theorem 1: *T is nowhere dense iff $\bar{T}$ is nowhere dense.*

Theorem 2: *A closed set T is nowhere dense iff T^c is dense.*

Theorem 3: *T is nowhere dense iff every non-empty open subset G of M possesses a non-empty open subset $\tilde{G}$ such that $T \cap \tilde{G} = \emptyset$.*

As trivial examples, we point out that any finite subset of R, and also the set of all integers, are nowhere dense subsets of R. A more interesting example is the following: Let M be the closed interval $[0, 1]$, with the metric obtained from R, and let us enumerate* the rational numbers in the interior of M. We thus obtain the sequence $r_1, r_2, r_3, \ldots$. Cover each

* We assume that the reader is acquainted with the fact that the set of rational numbers in any interval, or even in all of R, can be put in one-to-one correspondence with the natural numbers $1, 2, 3, \ldots$.

r_k with an open interval contained in the interior of M and having length $<(\frac{1}{10})^k$. The union S of these intervals is certainly a dense open subset of M, and so the complement, $M - S$, is nowhere dense. It is quite remarkable that the measure of S is less than $\frac{1}{10} + (\frac{1}{10})^2 + \cdots$, or $\frac{1}{9}$, so that $M - S$ possesses measure exceeding $\frac{8}{9}$. (The precise meaning of this sentence will become clear to the reader, if he is not yet acquainted with measure theory, in the following chapter.) Thus, although $M - S$, in some reasonable sense, consists of most of M, it is nowhere dense. In geometrical terms, the reason that $M - S$ is nowhere dense is that S, although having small measure, is formed from open intervals which are thoroughly scattered over the interval M.

The following theorem appears rather negative in spirit; it asserts that something cannot be done. Nevertheless, it is one of the most useful and powerful theorems in the theory of metric spaces. In the following section we shall give one extremely interesting application of this theorem, and it will also play a vital role in later developments.

Theorem 4 (Baire Category Theorem): *A complete metric space cannot be expressed as the union of countably many nowhere dense subsets.*

PROOF: Suppose that the complete metric space M can be expressed as $\bigcup_{n=1}^{\infty} A_n$, where each A_n is nowhere dense. Then we can find a non-empty open set G_1 which contains no point in common with A_1. Let x_1 be any point of G_1, and let ϵ_1 be chosen so small that $S_{\epsilon_1}[x_1] \subseteq G_1$. Since $S_{\epsilon_1}(x_1)$ is a non-empty open set we can find a non-empty open subset G_2 of $S_{\epsilon_1}(x_1)$ which is disjoint from A_2, and then we can choose a point x_2 in G_2 and a positive number ϵ_2 less than $\frac{1}{2}\epsilon_1$ and so small that $S_{\epsilon_2}[x_2] \subseteq G_2$. Continuing in this manner we obtain a sequence of nested closed balls $\{S_{\epsilon_k}[x_k]\}$, with ϵ_k approaching zero. By Exercise 4–12 the diameters of these balls also approach zero. Since M is complete, the Cantor intersection theorem assures us that there exists one (and only one) point contained in all the sets $S_{\epsilon_k}[x_k]$, and hence in *none* of the sets A_k. Thus, contrary to assumption, $M \neq \bigcup_{n=1}^{\infty} A_n$, and so the proof is complete.

The explanation of the rather undescriptive title of this theorem is the following: A metric space is said to be of *first category* if it *can* be expressed as a union of countably many nowhere dense sets; otherwise, it is of *second category*. Thus, the theorem which we have just proven can be stated in the following concise form: A complete metric space is of second category.

Exercises

1. Prove the assertion made in the second sentence of this section.

2. Prove the three theorems stated at the end of the second paragraph of this section.

3. Prove that the word "complete" may not be omitted in the statement of the Baire category theorem.

§8. THE EXISTENCE OF A CONTINUOUS NOWHERE-DIFFERENTIABLE FUNCTION

The functions that one usually encounters in elementary real analysis possess derivatives for all values of the independent variable, with the possible exception of a certain set of values which are rare in some sense. This led to the conviction during the early stages of the development of the differential calculus that a continuous function, except perhaps on a rare subset, is everywhere differentiable. We shall now show, by a suitable application of the Baire category theorem, that there exist continuous real-valued functions (defined on all of R) which are non-differentiable *everywhere*! It should be recalled, for clarity, that the function f is said to be *differentiable* at the point x of R if f is defined throughout some open interval containing x and if

$$\lim_{h\to 0, h\neq 0} \frac{f(x+h) - f(x)}{h}$$

exists *finitely*; equivalently, there exists a number α (necessarily unique) such that $|f(x+h) - f(x) - \alpha h| = o(h)$, where $o(h)$ denotes a function of h such that $o(h)/h \to 0$ as $h \to 0$.

Let P be the collection of all real-valued continuous functions defined on R and possessing period one. If we define the distance between two members of P by the equation

$$\rho(f, g) = \max_{x\in R} |f(x) - g(x)| \left(= \max_{x\in[0,1]} |f(x) - g(x)|\right),$$

P clearly becomes a *complete* metric space. We shall find it convenient to employ the notation $\|f\|$ instead of $\rho(f, 0)$ (where 0 here denotes the function which vanishes identically on R), so that $\|f - g\| = \rho(f, g)$. We shall use the following two facts about P:

(*a*) Given any positive number ϵ (however small) and any positive number k (however large), there exists in P a member f which satisfies the inequality $\|f\| < \epsilon$ and which possesses everywhere a left-hand and a right-hand derivative which are *finite* but greater in absolute value than k. (Remember that if f possesses at x finite left-hand and right-hand derivatives, then f is differentiable at x iff the values of the left-hand and right-hand derivatives coincide.) Rather than giving a formal proof of this assertion, we suggest that the reader think of a saw whose cutting edge consists of many short but steep teeth.

(*b*) Given any f in P and any positive number ϵ, there exists in P a function g which is continuously differentiable everywhere in R and whose distance from f is less than ϵ. (That is, the continuously differentiable members of P form a dense subset of P.) Without going into all the details, we sketch the proof as follows: From a basic theorem of real analysis, the function f, being continuous, can be approximated within $\epsilon/3$ everywhere on $[0, 1]$ by a polygonal function g_1, a continuous function whose graph in this interval consists of a finite number of straight line segments. If $g_1(0) \neq g_1(1)$, it is possible to find another polygonal function g_2 such that $g_2(0) = g_2(1)$ and such that $|g_2(x) - g_1(x)| < \epsilon/3$, and hence $|g_2(x) - f(x)| < 2\epsilon/3$, everywhere on $[0, 1]$. The latter inequality persists if g_2 is defined everywhere on R by the periodicity condition $g_2(x + 1) = g_2(x)$. Finally, the corners appearing in the graph of g_2 can be rounded off to furnish a periodic continuously differentiable function g such that $|g(x) - g_2(x)| < \epsilon/3$, and hence $|g(x) - f(x)| < \epsilon$, everywhere.

Now for every positive integer n we define a subset E_n of P in the following manner: The function f (belonging to P) is contained in E_n iff there exists *at least* one value of x such that $|(f(x + h) - f(x))/h| \leqslant n$ for all h in the open interval $(0, 1/n)$. Note carefully that if f has a *finite* right-hand derivative f'_+ *at even one value of* x, f belongs to some E_n. (For example, if $f'_+(x_0) = -27.3$, then $|(f(x_0 + h) - f(x_0))/h| \leqslant 28$ for all sufficiently small positive h—not necessarily for $0 < h < \frac{1}{28}$, but, let us say, for $0 < h < \frac{1}{973}$. Then clearly $f \in E_{973}$.) In order not to break the main line of the argument, let us accept momentarily the fact that each E_n is a *closed* and *nowhere dense* subset of P. Since P is complete, $\bigcup_{n=1}^{\infty} E_n \neq P$ (by the Baire category theorem), and so there exists an element of P (that is, a periodic continuous function) which is *outside* every E_n. This function does not have a finite right-hand derivative *anywhere*, and therefore it is certainly nowhere differentiable.

It now remains to justify the previous assertions that each E_n is closed and that each E_n is nowhere dense. (Actually, only the second assertion is needed in order to justify the preceding proof, but we shall use the first in proving the second.)

(*a*) Suppose that E_n is *not* closed. Then there exists a member f of P which is *not* in E_n and there exists a sequence $f_1, f_2, \ldots$ of members of E_n such that $\|f - f_k\| \to 0$ as $k \to +\infty$. For each k we can choose a point x_k somewhere in $[0, 1]$ such that $|(f_k(x_k + h) - f_k(x_k))/h| \leqslant n$ whenever $0 < h < 1/n$. By the Bolzano-Weierstrass theorem we can select a subsequence of the f_k's in such a manner that the corresponding x_k's converge to some point, ξ, in $[0, 1]$. Call this subsequence of functions $\tilde{f}_1, \tilde{f}_2, \ldots$ and the corresponding subsequence of the x_k's $\xi_1, \xi_2, \ldots$. Then $\|f - \tilde{f}_k\| \to 0$ and $\xi_k \to \xi$ as $k \to +\infty$. Choose any number h in

$(0, 1/n)$. Then, for any index k,

$$\left|\frac{f(\xi+h)-f(\xi)}{h}\right| = \left|\frac{f(\xi+h)-f(\xi_k+h)}{h} + \frac{f(\xi_k+h)-\tilde{f}_k(\xi_k+h)}{h} + \frac{\tilde{f}_k(\xi_k+h)-\tilde{f}_k(\xi_k)}{h} + \frac{\tilde{f}_k(\xi_k)-f(\xi_k)}{h} + \frac{f(\xi_k)-f(\xi)}{h}\right|$$

$$\leqslant \frac{|f(\xi+h)-f(\xi_k+h)|}{h} + \frac{|f(\xi_k+h)-\tilde{f}_k(\xi_k+h)|}{h} + \frac{|\tilde{f}_k(\xi_k+h)-\tilde{f}_k(\xi_k)|}{h} + \frac{|\tilde{f}_k(\xi_k)-f(\xi_k)|}{h} + \frac{|f(\xi_k)-f(\xi)|}{h}.$$

Now, the second and fourth numerators are each $\leqslant \|f-\tilde{f}_k\|$ and the third term is $\leqslant n$. We can, given any $\epsilon > 0$, choose an index K so large that, whenever $k > K$, $\|f-\tilde{f}_k\| < \epsilon h$. Also, f is *uniformly* continuous (why?), and so, given $\epsilon > 0$, we can choose a positive number δ such that $|f(t)-f(\tau)| < \epsilon h$ whenever $|\tau - t| < \delta$. Now let us choose $\tilde{K}$ so large that $|\xi - \xi_k| < \delta$ whenever $k > \tilde{K}$. (Note that $|(\xi+h)-(\xi_k+h)| = |\xi - \xi_k|$.) Then, for $k > \max\{K, \tilde{K}\}$ the first, second, fourth, and fifth numerators are each less than ϵh, and so the entire right-hand side is less than $n + 4\epsilon$. Now the *left* side has nothing to do with k or ϵ, and so we obtain $|(f(\xi+h)-f(\xi))/h| \leqslant n$ for every h in $(0, 1/n)$. Thus, $f \in E_n$, contrary to the hypothesis $f \notin E_n$. Therefore, in fact, E_n *is* closed.

(*b*) Suppose $f \in E_n$. We shall show that, for any $\epsilon > 0$, there exists a member of P, say g, such that $g \notin E_n$ and $\|f-g\| < \epsilon$. This (taking account of the fact that E_n is closed) shows that E_n is nowhere dense. By the auxiliary result (*b*) presented earlier, we can find a *continuously differentiable* member of P (call it h) such that $\|f-h\| < \epsilon/2$. By the auxiliary result (*a*), there exists a member of P (call it $\tilde{h}$) such that $\|\tilde{h}\| < \epsilon/2$ and $|\tilde{h}'_+| > n + \max|h'|$ everywhere. Then let $g = h - \tilde{h}$. By the triangle inequality, $\|f-g\| = \|(f-h)+\tilde{h}\| \leqslant \|f-h\| + \|\tilde{h}\| < \epsilon/2 + \epsilon/2 = \epsilon$. Also, g'_+ exists everywhere, $g'_+ = h' - \tilde{h}'_+$, and $|g'_+| \geqslant |\tilde{h}'_+| - |h'| > n + \{\max|h'| - |h'|\} \geqslant n$. Thus, g does not belong to E_n, and so E_n is nowhere dense.

It should be emphasized that the preceding proof is existential, or non-constructive. Actually, the first proof of the existence of a continuous nowhere-differentiable function was constructive, for Weierstrass demonstrated that the function defined by the convergent series

$$f(x) = \sum_{k=0}^{\infty} b^k \cos a^k x \qquad (0 < b < 1,\ ab > 1 + \tfrac{3}{2}\pi)$$

possesses this property. Since Weierstrass made his striking discovery, a succession of simpler examples have been found. Perhaps the simplest known example (due to van der Waerden) is the function defined by the series

$$\sum_{k=1}^{\infty} \frac{[10^k x]}{10^k},$$

where, for any real number w, $[w]$ denotes the absolute value of the difference between w and the integer closest to w. Both of these examples are discussed in [Titchmarsh, pp. 351–354].

§9. CONTINUOUS FUNCTIONS

Let f be a real-valued function defined on R. The reader will recall that f is said to be *continuous at* x_0 (where $x_0 \in R$) if for every positive number ϵ there exists a positive number δ such that $|f(x) - f(x_0)| < \epsilon$ whenever $|x - x_0| < \delta$. Furthermore, the function f is said to be *continuous* if it is continuous at each point in R. (These definitions are easily modified to apply to the case when the domain of the function f is a proper non-empty subset of R, but we shall not concern ourselves with this matter.) The reader should also be familiar with the sequential definition of continuity: The function f is *continuous at* x_0 if for every sequence of real numbers $x_1, x_2, x_3, \ldots$ converging to x_0, the corresponding sequence $f(x_1), f(x_2), f(x_3), \ldots$ converges to $f(x_0)$. (Of course, it is a *theorem* of real analysis that the epsilon-delta definition and the sequential definition of continuity are equivalent.) Now, it is a simple matter to generalize both definitions of continuity to a function mapping a metric space M into a metric space N; the reader should have no difficulty in seeing that the proof of the equivalence of these two definitions is substantially the same as the proof in the classical case just referred to.

Definition 1: *The function f mapping M into N (where, of course, M and N may be the same metric space) is said to be continuous at the point x_0 of M if for every positive number ϵ there exists a positive number δ such that $\rho(f(x), f(x_0)) < \epsilon$ whenever $\rho(x, x_0) < \delta$. (Of course, the two appearances of the symbol ρ refer to the metrics on N and M, respectively.) Equivalently, the function f is said to be continuous at the point x_0 of M if for every sequence $x_1, x_2, x_3, \ldots$ in M which converges to x_0 the corresponding sequence $f(x_1), f(x_2), f(x_3), \ldots$ in N converges to $f(x_0)$. The function f is said to be continuous if it is continuous at all points of M.*

It is a most significant and interesting fact that the two given equivalent definitions of continuity are, in turn, equivalent to a definition in which only the concept of an open set is involved. (This fact is employed to provide the definition of continuity when mappings between *topological* spaces are considered.)

Theorem 1: *The function f mapping the metric space M into the metric space N is continuous at the point x_0 (of M) iff the inverse image* of every neighborhood† of $f(x_0)$ contains a neighborhood of x_0. The function f is continuous iff the inverse image of every open subset of N is an open subset of M.*

PROOF: Suppose f is continuous at x_0. Then, according to the preceding definition, the inverse image of the open ball $S_\epsilon(f(x_0))$ contains the open ball $S_\delta(x_0)$, where ϵ is any given positive number and the positive number δ is chosen appropriately. Since every neighborhood of $f(x_0)$ contains some open ball $S_\epsilon(f(x_0))$, the inverse image of any neighborhood of $f(x_0)$ must contain a neighborhood of x_0. Conversely, suppose that the inverse image of every neighborhood of $f(x_0)$ contains a neighborhood of x_0; then, in particular, the inverse image of $S_\epsilon(f(x_0))$ contains a neighborhood of x_0, and a neighborhood of x_0 must contain $S_\delta(x_0)$ for sufficiently small δ. This completes the proof of the first part of the theorem.

The proof of the second half is an easy extension of the foregoing argument, and is left to the reader as Exercise 1.

It should be emphasized that for a given ϵ the choice of δ usually depends on the point x_0; in general it is not possible to choose a single δ which will suffice at each point of M. This remark leads to the following definition.

Definition 2: *The function f (mapping M into N) is said to be uniformly continuous if for every positive number ϵ it is possible to find a number δ (depending only on ϵ and the function f) such that $\rho(f(x), f(x_0)) < \epsilon$ whenever $\rho(x, x_0) < \delta$.*

---

* If f is a function mapping M into N (where M and N are arbitrary non-empty sets, not necessarily metric spaces) and if A is any subset of M, then $f(A)$, the *image* of A (under the function f) is the subset of N consisting of all objects y such that $y = f(x)$ for at least one member x of A. Conversely, if B is any subset of N, then $f^{-1}(B)$, the *inverse image* of B, is the subset of M consisting of all objects x in M such that $f(x) \in B$. Exercise 6 lists the most important facts about images and inverse images.

† A *neighborhood of a point* x_0 in a metric space M is an open subset of M which contains x_0. Some authors use the term "neighborhood of x_0" to mean a subset A of M, not necessarily open, such that there exists an open set G satisfying the conditions $x_0 \in G \subseteq A$.

It is one of the most important theorems of real analysis that a continuous real-valued function defined on a bounded closed subset of R is uniformly continuous. In §10 we shall discuss certain ideas which enable us to generalize this result. (Cf. Exercise 10–5.)

Exercises

1. Complete the proof of Theorem 1.

2. (*a*) Let $f(x) = x^2$ for every real number x. Show that the function f is continuous, but not uniformly continuous. On the other hand, show that the restriction of f to any bounded interval (open, closed, or half-open) is indeed uniformly continuous.
(*b*) Let $f(x) = x/(1 + x^2)$ for every real number x. Show that f is uniformly continuous.

3. Let f_1 and f_2 be continuous real-valued functions defined on the metric space M, and let α and β be any real numbers. Prove that the function $\alpha f_1 + \beta f_2$ is also continuous, and that if f_1 and f_2 are both uniformly continuous, $\alpha f_1 + \beta f_2$ is also uniformly continuous. Also, prove that the product $f_1 f_2$ is continuous and that if f_2 never assumes the value zero the quotient f_1/f_2 is also continuous.

4. (*a*) Let $f_1, f_2, f_3, \ldots$ be a sequence of continuous functions mapping the metric space M into the metric space N, and suppose that $\lim_{n\to\infty} f_n(x)$ exists for every point x of M. Show by an example that the limit function may be discontinuous.
(*b*) Suppose that the sequence appearing in part (*a*) is uniformly convergent—i.e., that for every positive number ϵ there exists an integer $N(\epsilon)$, independent of x, such that $\rho(f_k(x), \lim_{n\to\infty} f_n(x)) < \epsilon$ whenever $k > N(\epsilon)$, for all points x in M. Prove that in this case the limit function is certainly continuous.
(*c*) Demonstrate by means of an example that the limit of the sequence $f_1, f_2, f_3, \ldots$ of continuous functions may be continuous even though the sequence may fail to converge uniformly. (For convenience, consider functions mapping R into R, i.e., choose $M = N = R$.)

5. (*a*) Let M be a metric space, $\tilde{M}$ a non-empty subset of M, and x a point of M. Then $\rho(x, \tilde{M})$, the distance from x to $\tilde{M}$, is defined as $\inf_{y\in\tilde{M}} \rho(x, y)$. Prove that $\rho(x, \tilde{M}) = 0$ iff $x \in \bar{\tilde{M}}$. (Recall that $\bar{\tilde{M}}$ denotes the closure of $\tilde{M}$.) Also show that $\rho(x, \tilde{M})$ is, for fixed $\tilde{M}$, a continuous function from M into R.
(*b*) Let M_1 and M_2 be non-empty, closed, disjoint subsets of the metric space M. Prove that there exists a continuous real-valued

function f defined on M which satisfies the following conditions: $f(x) = 0$ if $x \in M_1$, $f(x) = 1$ if $x \in M_2$, $0 < f(x) < 1$ if $x \notin M_1 \cup M_2$. (A similar theorem holds true in topological spaces possessing the property called normality; this theorem, known as Urysohn's lemma, is one of the most important results in general topology.)

(*c*) Let M_1 and M_2 be non-empty subsets of a metric space. The distance between these subsets, denoted $\rho(M_1, M_2)$, is defined as $\inf_{x \in M_1, y \in M_2} \rho(x, y)$. (Obviously, if M_1 and M_2 are not disjoint the distance between them is zero.) Show that there exist two non-empty, closed, disjoint subsets, M_1 and M_2, of R such that $\rho(M_1, M_2) = 0$.

6. Let f be a function mapping the set M into the set N, let $\{A_\alpha\}$ denote any collection of subsets of M, and let $\{B_\beta\}$ denote any collection of subsets of N. Prove the following equalities and inclusions:

(*i*) If $A_{\alpha_1} \subseteq A_{\alpha_2}$, then $f(A_{\alpha_1}) \subseteq f(A_{\alpha_2})$; $f(\bigcup A_\alpha) = \bigcup f(A_\alpha)$; $f(\bigcap A_\alpha) \subseteq \bigcap f(A_\alpha)$;

(*ii*) If $B_{\beta_1} \subseteq B_{\beta_2}$, then $f^{-1}(B_{\beta_1}) \subseteq f^{-1}(B_{\beta_2})$; $f^{-1}(\bigcup B_\beta) = \bigcup f^{-1}(B_\beta)$; $f^{-1}(\bigcap B_\beta) = \bigcap f^{-1}(B_\beta)$.

Show by a simple example that the inclusion relation appearing in the last part of (*i*) cannot be replaced by equality.

§10. COMPACTNESS

The property of compactness is one of the most significant that arises in the study of metric spaces (and, more generally, in the study of topological spaces). We shall discuss this important concept only very briefly in this section, leaving the proof of the most important result to the reader in the form of a chain of exercises.

We begin by recalling the Heine-Borel theorem, which should be familiar to the reader from his study of the fundamentals of real analysis:

Let A be a non-empty, bounded, closed subset of R, and let $\{G_\alpha\}$ be a collection of open subsets of R such that $A \subseteq \bigcup G_\alpha$. Then it is possible to select a finite number of G_α's, say $G_{\alpha_1}, G_{\alpha_2}, \ldots, G_{\alpha_n}$, such that $A \subseteq \bigcup_{k=1}^{n} G_{\alpha_k}$. Also, from every sequence $x_1, x_2, x_3, \ldots$ in A it is possible to extract a subsequence which converges to a member of A. Conversely, if either of the preceding conclusions holds true, so does the other, and A must be closed and bounded.*

* Strictly speaking, the Heine-Borel theorem consists only of the first two sentences of this paragraph. The third sentence follows easily from the Bolzano-Weierstrass theorem and the definition of a closed set, while the proof of the last sentence is quite elementary. We remind the reader of the fact that the Heine-Borel theorem is the essential tool needed in proving (among other things) that a continuous real-valued (or complex-valued) function defined on a bounded closed subset of R is uniformly continuous.

We now proceed to present a number of definitions which provide the basis for suitably extending the Heine-Borel theorem to a certain class of metric spaces.

Definition 1: *An open covering of a metric space M is a collection* $\{G_\alpha\}$ *of open subsets of M whose union is M; that is, each point of M is contained in at least one of the* G_α*'s.*

Definition 2: *A metric space M is said to be compact if from every open covering* $\{G_\alpha\}$ *of M it is possible to extract a finite subcollection of the* G_α*'s which constitute an open covering of M. A subset N of the metric space M is said to be compact if N is either empty or is compact when considered as a metric space in its own right (with the metric derived from M).*

Definition 3: *A metric space is said to be locally compact if each point has a neighborhood whose closure is compact.*

Definition 4: *A metric space M is said to be sequentially compact if from every sequence* $x_1, x_2, x_3, \ldots$ *in M it is possible to extract a subsequence which is convergent.*

Definition 5: *A B-W (Bolzano-Weierstrass) space is a metric space having the property that every infinite subset has at least one limit-point (which need not belong to the subset).*

Definition 6: *A metric space M is said to be totally bounded if for every positive number* ϵ *there exists a finite collection of points*

$$\{x_1, x_2, \ldots, x_n\}$$

such that the open balls $\{S_\epsilon(x_1), S_\epsilon(x_2), \ldots, S_\epsilon(x_n)\}$ *constitute an open covering of M.*

Theorem 1: *If a metric space possesses any one of the following four properties, it possesses the other three, so that all four properties are equivalent:*

(*a*) *compactness,*
(*b*) *sequential compactness,*
(*c*) *B-W property,*
(*d*) *completeness and total boundedness.*

The proof is contained in Exercises* 9 to 15. The reader will certainly

* Although the reader is strongly urged to try to prove these results without referring to outside sources, it may be mentioned that an excellent presentation of the proof of Theorem 1 will be found in [Simmons, pp. 121–125].

find it helpful to keep in mind, while working these problems, the proof from real analysis that a subset of R is compact iff it is closed and bounded.

Now let M be a compact metric space and let $C(M)$ be the class of all continuous real-valued functions defined on M. From Exercises 9–3 and 10–4 it is easily seen that $C(M)$ becomes a metric space if the distance between any two members f and g of $C(M)$ is defined by the equation $\rho(f, g) = \max_{x \in M} |f(x) - g(x)|$, and from Exercise 9–4 it becomes evident that $C(M)$, when provided with this metric, is complete. Clearly $C(M)$ is not compact, for it is not bounded* (simply consider the constant functions!), and hence certainly not totally bounded (cf. Exercise 3). This remark suggests the problem of characterizing the compact subsets of $C(M)$. By referring to condition (*d*) of Theorem 1 and recalling (cf. Exercise 4–2) that a (non-empty) subset of a complete metric space is complete iff it is closed, we see that it remains only to obtain a suitable description of total boundedness in $C(M)$. For this purpose we introduce the concept of equicontinuity.

Definition 7: *A (non-empty) collection of functions $\{f_\alpha\}$ from a metric space M into a metric space N is said to be equicontinuous if for every positive number ϵ there exists a positive number δ such that, for every index α, $\rho(f_\alpha(x), f_\alpha(y)) < \epsilon$ whenever $\rho(x, y) < \delta$.*

The following theorem is now an easy consequence of the preceding definition and the preceding theorem.

Theorem 2 (Ascoli-Arzela): *A (non-empty) subset of $C(M)$, where M is a compact metric space, is compact iff it is closed, bounded, and equicontinuous.*

Exercises

1. Show that a compact subset of a metric space must be closed and that a closed subset of a compact metric space is compact.

2. Show that the metric space defined in Example (*b*) of §2 is bounded, but that it is not totally bounded if it contains infinitely many points. Show directly that in the latter case the space is not compact.

3. Show that a totally bounded metric space is bounded. (The preceding exercise shows that the converse is not true.)

4. Let f be a continuous function mapping the metric space M into the metric space N. Show that if M' is a compact subset of M, then $f(M')$ is compact. (In particular, by choosing $N = R$,

---

* A metric space is said to be *bounded* if its diameter is finite.

we see that a continuous real-valued function defined on a compact metric space is bounded and attains its sup and inf.)

5. Prove that a continuous function from a compact metric space into a metric space (not necessarily compact) is uniformly continuous.

6. Show that, although each of the functions $\sin nx$, $n = 1, 2, 3, \ldots$, is uniformly continuous on the interval $[0, 2\pi]$, the collection of all these functions is not equicontinuous.

7. Let m be a positive constant, and let S_m be the subset of $C([a, b])$ consisting of all functions possessing at each point of $[a, b]$ a derivative not exceeding m in absolute value. Show that this collection of functions is equicontinuous.

8. Prove directly (i.e., by using standard techniques of real analysis) the following particular case of Theorem 2: Let $\{f_\alpha\}$ be an infinite equicontinuous collection of real-valued functions defined on the interval $[a, b]$. From every uniformly bounded sequence drawn from this collection it is possible to extract a subsequence which converges uniformly on the specified interval. Furthermore, the condition of uniform boundedness may be replaced by the (seemingly) weaker condition that the chosen sequence should be bounded at any one point of the interval.

9. Suppose that the metric space M contains an infinite subset M' from which it is impossible to extract a convergence subsequence. Show that each point of M' can be covered by an open set which contains no other point of M' and that each point of $M - M'$ can be covered by an open set which contains no point of M', and deduce from this result that compactness implies sequential compactness.

10. Prove, with the aid of the preceding exercise, that sequential compactness is equivalent to the *B-W* property. (Hint: Recall the proof of the Bolzano-Weierstrass theorem on the real line.)

11. Prove that compactness implies the *B-W* property (and hence, by the preceding exercise, sequential compactness).

12. Let the collection $\{G_\alpha\}$ constitute an open covering of a sequentially compact metric space. Show that there exists a positive number δ (called a Lebesgue number of the collection $\{G_\alpha\}$) such that whenever $\rho(x, y) < \delta$ there exists at least one G_α which contains both x and y.

13. Use the preceding result to show that sequential compactness implies total boundedness.

14. Prove that sequential compactness implies compactness.

15. Prove that compactness is equivalent to completeness together with total boundedness.

16. Let $f_1, f_2, f_3, \ldots$ be a sequence of continuous real-valued functions defined on a compact space M, suppose that the inequalities $f_1(x) \leqslant f_2(x) \leqslant f_3(x) \leqslant \cdots$ hold at every point x of M, and suppose that this sequence of functions converges (everywhere on M) to a *continuous* function. Prove that the convergence is uniform. (This is Dini's theorem of monotone convergence.)

CHAPTER 2

LEBESGUE MEASURE AND INTEGRATION

§1. INTRODUCTORY REMARKS

Almost all significant applications of the theory of linear spaces involve to a greater or lesser extent the concept of integration. The Riemann integral, useful though it is for many purposes, has proven to be quite inadequate in many problems of a somewhat sophisticated nature, and it has been largely superseded by the Lebesgue integral and extensions of the latter.

The deficiencies of the Riemann integral can be roughly summed up in two brief statements: first, not enough functions are integrable; second, and more seriously, limiting operations often lead to insurmountable difficulties. In particular, if the functions $f_1, f_2, \ldots$ are each integrable on a bounded closed interval $[a, b]$ and if they converge everywhere in this interval to a function f, it is *not* always true that

$$\lim_{n \to \infty} \int_a^b f_n(x)\, dx = \int_a^b f(x)\, dx.$$

Three things can go wrong: (*a*) the limit on the left side may not exist; (*b*) even if this limit does exist, the function f may not be integrable, so that the right side may be meaningless; (*c*) even if both sides of the preceding equality exist, they may nevertheless be unequal. (Cf. Exercise 1.)

In the early 1900's H. Lebesgue presented a new theory which eliminates most of the deficiencies of the Riemann theory, and it has had a tremendous impact on the development of mathematics since that time. In this chapter we present a limited portion of Lebesgue's theory, and in later chapters we shall see how this theory plays an important role in many parts of linear analysis.

There are many ways to present Lebesgue's theory, and it is not always easy to show that two different developments are actually equivalent. Roughly speaking, the various approaches fall into two main groups. In one group are the approaches in which *measure* comes first and *integration* comes second; in the second group the order is reversed. We shall present an approach of the first kind.

Although vast generalizations of Lebesgue's theory have been made, we shall confine attention in most of this chapter to the case that all sets under consideration are subsets of the real line R. (In the closing sections we shall indicate briefly how the Lebesgue theory is extended to integration over subsets of the plane and to integration over more abstract sets.) In order to avoid endless repetition of the word "open," we make the convention that any set named O (with or without a subscript, such as O_1, O_2, O_k, and so forth) is open (in terms of the metric of R). It must constantly be kept in mind that any open subset of R can be expressed *in a unique manner* (except for order) as the *disjoint* union of a finite or countably infinite collection of open intervals. These intervals are termed the *components* of the given set. (An interval need not be of finite extent; for example, the unique representation of $R - \{3\}$ as a disjoint union of open intervals is $(-\infty, 3) \cup (3, +\infty)$.) For the benefit of the reader who may be unacquainted with the preceding fact concerning the structure of open subsets of R, we present in Appendix E a brief proof.

We shall find it necessary to deal with infinite series whose terms are either positive numbers or $+\infty$; since the reader may not be at ease with the extended real number system, we present in Appendix F a brief review of the theory of convergent series of real numbers and an indication of how the pertinent ideas have to be modified when one introduces the *ideal number* $+\infty$. The essential point for our purposes is the following fact: To every series $a_1 + a_2 + a_3 + \cdots$, where each of the a's is a positive real number or $+\infty$, we can associate a sum which is also a positive real number or $+\infty$, and this sum is unaffected by any rearrangement of the order of the terms.

Exercise

1. Illustrate each of the three unpleasant possibilities cited in the second paragraph.

§2. MEASURE OF OPEN SETS

Measure (more precisely, Lebesgue measure) is an extension of the concept of length. We begin by defining, in an obvious manner, the measure of open sets, and then we shall prove certain properties of the measure. (While the *definition* is obvious, the proofs of certain *seemingly* obvious properties require a considerable skill.) In the following section we shall extend the concept of measure and the proofs of its properties to sets which are not necessarily open, and then in the remaining sections of the main part of the chapter we shall define the Lebesgue integral and establish its most important properties.

Before undertaking the systematic study of the theory developed in the present and following sections, the reader may find it possible to get an overall feeling for the subject by rapidly reading the definitions and the theorems several times. Hopefully, he will then see the general pattern of development and will not find the technical details such a stumbling-block as he might otherwise.

Definition 1: *The measure of O, denoted $\mu(O)$, is the sum of the lengths of the components of O. (For example, $\mu(R) = +\infty$, $\mu((0, +\infty)) = +\infty$, $\mu((1, 3)) = 2$, $\mu((1, 3) \cup (7, 11)) = 6$, $\mu(\emptyset) = 0$.)*

Note that this definition makes sense, because of the *unique* decomposition of O into components and the fact that the sum of the lengths of the components does not depend on the order of summation. Note also that if O is an open interval, $\mu(O)$ coincides with the length of O. Thus, measure is indeed a generalization of length.

Theorem 1: *If $O_1 \subseteq O_2$, then $\mu(O_1) \leqslant \mu(O_2)$.*

PROOF: Break up O_2 into its components, $C_1, C_2, \ldots$. Each component of O_1 is clearly entirely contained in one of the C_k's. Thus, for each C_k we can determine the disjoint collection (perhaps empty, at most countably infinite) of components of O_1 contained in C_k; we denote these components $\Gamma_{1,k}, \Gamma_{2,k}, \ldots$. If, for a particular k, the corresponding collection of Γ's is empty or finite, then clearly

$$\sum_j \mu(\Gamma_{j,k}) \leqslant \mu(C_k).$$

If C_k contains infinitely many components of O_1, the preceding inequality still holds, as is seen by enumerating these components, forming finite sums,* and passing to the limit. If we now sum the preceding inequality over all values of k, we have on the left the sum of the lengths, in some order, of all components of O_1, while on the right we have the sum of the lengths, in some order, of all components of O_2. Hence we have obtained the desired result, $\mu(O_1) \leqslant \mu(O_2)$.

Theorem 2: *If $O_1, O_2, \ldots$ are disjoint,† then $\mu(O_1 \cup O_2 \cup \cdots) = \mu(O_1) + \mu(O_2) + \cdots$.*

PROOF: Since the O_k's are disjoint, it is immediately seen that the components of $O_1 \cup O_2 \cup \cdots$ are simply the components of O_1, the components of $O_2, \ldots$, *without duplication.* Thus the sum on the right is simply a rearrangement of the sum on the left, and so the equality must hold.

Theorem 3: *Whether or not the sets $O_1, O_2, \ldots$ overlap,*

$$\mu(O_1 \cup O_2 \cup \cdots) \leqslant \mu(O_1) + \mu(O_2) + \cdots.$$

PROOF: If the right side of the inequality is $+\infty$, there is nothing to prove. Therefore, we may assume that $\mu(O_1) + \mu(O_2) + \cdots$ is finite. Let $C_1, C_2, \ldots$ be the components of $O_1 \cup O_2 \cup \cdots$, and let $\Gamma_{i,k}$ denote $O_i \cap C_k$. Suppose that we can prove the inequality

$$\mu(\Gamma_{1,k}) + \mu(\Gamma_{2,k}) + \cdots \geqslant \mu(C_k). \tag{2–1}$$

Then by summing over k we would obtain

$$\sum_k \{\mu(\Gamma_{1,k}) + \mu(\Gamma_{2,k}) + \cdots\} \geqslant \sum_k \mu(C_k). \tag{2–2}$$

A moment's thought will show that the left side of (2–2) is the summation, in some order, of the lengths of the individual components of the individual O_k's, while on the right side we have the summation, in some order, of the lengths of the components of $O_1 \cup O_2 \cup \cdots$. Thus (2–2) is equivalent to the assertion of the theorem. However, the inequality (2–1), which seems obvious, but isn't, must be established. (See the Remark at the end of the proof.)

---

* Here we employ the following fact: If the intervals $J_1, J_2, \ldots, J_n$ are disjoint and are all contained in the interval I, then the sum of the lengths of the J's cannot exceed the length of I. It clearly does not matter whether the intervals are open, closed, or half-open.

† Here and elsewhere the notation $\cdots$ implies that we are dealing with either a finite or countably infinite collection of sets. Also, keep in mind that expressions such as $\mu(O_1 \cup O_2 \cup \cdots)$ are meaningful, since any union of open sets is open.

Choose a particular C_k; it is of finite length or infinite length. We shall consider only the former case, but the case of infinite length is disposed of by a simple extension of the argument which we now present. (In fact, it will become apparent, when the argument that is about to be presented is extended to the case of an interval of infinite length, that no such interval can exist when the hypothesis made in the second sentence of this proof is satisfied. Of course, this is exactly what one would expect.) Now suppose that (2–1) is false—that is, suppose that

$$\mu(\Gamma_{1,k}) + \mu(\Gamma_{2,k}) + \cdots < \mu(C_k). \tag{2–3}$$

From the definition of the C's and the Γ's (a drawing may be helpful), it is evident that

$$\Gamma_{1,k} \cup \Gamma_{2,k} \cup \cdots = C_k. \tag{2–4}$$

Since C_k is an open interval of finite length, we may write $C_k = (a, b)$, so that (2–3) assumes the form

$$\mu(\Gamma_{1,k}) + \mu(\Gamma_{2,k}) + \cdots < b - a. \tag{2–5}$$

We can find a number a' slightly larger than a and a number b' slightly smaller than b so that

$$\mu(\Gamma_{1,k}) + \mu(\Gamma_{2,k}) + \cdots < b' - a'. \tag{2–6}$$

However, from (2–4) we observe that $\Gamma_{1,k} \cup \Gamma_{2,k} \cup \cdots \supset [a', b']$, and it now follows from the Heine-Borel theorem that a *finite* number of open sets $\Gamma_{1,k}, \Gamma_{2,k}, \ldots$ cover $[a', b']$. Now, it is *really* obvious (and true!) that the sum of the lengths of this *finite* collection of Γ's must exceed $b' - a'$; and since this *finite* sum cannot exceed the left side of (2–6), we obtain

$$\mu(\Gamma_{1,k}) + \mu(\Gamma_{2,k}) + \cdots > b' - a'. \tag{2–7}$$

Since (2–6) and (2–7) contradict each other, we conclude that (2–1) is true, and so the proof is complete.

Remark: The subtlety that is often misunderstood or overlooked is the following: Might it be possible to cover an interval I by a countable collection of intervals $\{I_1, I_2, \ldots\}$ whose lengths add up to *less than* the length of I? The preceding argument shows that the answer to this question is *no*, but observe that the proof requires use of the Heine-Borel theorem, and there is no way of avoiding it (or something equivalent to it). Thus, in this case, common sense furnishes the correct answer, but the reader hopefully realizes that common sense is often misleading. Exercise 1 should help the reader gain a deeper understanding of the fact that the theorem which we have proven is very far from obvious.

Theorem 4: $\mu(O_1 \cup O_2) + \mu(O_1 \cap O_2) = \mu(O_1) + \mu(O_2)$. *(Roughly speaking, this asserts that when we add the measures of O_1 and O_2 we count the overlapping part twice. The reader may find it helpful to think of O_1 and O_2 as clubs and $\mu(O_1)$ and $\mu(O_2)$ as the number of members.)*

PROOF: Since this theorem is trivially true if either $\mu(O_1)$ or $\mu(O_2)$ is $+\infty$ (cf. Theorem 1), we may assume that $\mu(O_1)$ and $\mu(O_2)$ are both finite.

(*a*) If O_1 and O_2 are disjoint, Theorem 2 settles the matter.

(*b*) Suppose that O_1 and O_2 are not disjoint, but suppose that O_1 and O_2 each have a finite number of components and that each component of O_1 is either *disjoint* from O_2 or *coincides* with a component of O_2, and vice-versa. (For example, let $O_1 = (1, 2) \cup (3, 5)$ and let $O_2 = (-7, 1) \cup (3, 5) \cup (8, 11)$.) Then

$$\mu(O_1) = \sum \mu(\text{shared components}) + \sum \mu(\text{unshared components})$$

and

$$\mu(O_2) = \sum \mu(\text{shared components}) + \sum \mu(\text{unshared components}).$$

Adding, we clearly obtain

$$\mu(O_1) + \mu(O_2) \\ = \{\sum \mu(\text{shared components}) + \sum \mu(\text{all unshared components})\} \\ + \sum \mu(\text{shared components}).$$

The quantity in { } is clearly $\mu(O_1 \cup O_2)$ and the last sum is clearly $\mu(O_1 \cap O_2)$; thus we have the desired equality.

(*c*) Now we still assume that O_1 and O_2 overlap and that each of them has a finite number of components, but we drop the restriction imposed in (*b*) concerning the nature of the overlap. This means (a simple drawing may help) that the components of O_1 are permitted to have endpoints which are interior points of O_2, and conversely. Now simply remove from O_1 those points (if any) which are endpoints of components of O_2, and conversely. We then obtain sets $\tilde{O}_1$ and $\tilde{O}_2$ which are clearly open (since only a finite number of points are removed), and these new sets satisfy the conditions imposed in (*b*). Thus

$$\mu(\tilde{O}_1) + \mu(\tilde{O}_2) = \mu(\tilde{O}_1 \cup \tilde{O}_2) + \mu(\tilde{O}_1 \cap \tilde{O}_2). \qquad (2\text{–}8)$$

From the simple relationship between the sets O_1, O_2, $\tilde{O}_1$, $\tilde{O}_2$ it is clear that $\tilde{O}_1$ and $\tilde{O}_2$ may be replaced in (2–8) by O_1 and O_2, respectively, and so we have obtained once again the desired equality.

(*d*) Finally, suppose that at least one of the two given sets has infinitely many components. Given any $\epsilon > 0$, split O_1 into $A_1 \cup R_1$ and O_2 into $A_2 \cup R_2$, where A_1 and A_2 consist of a *finite* number of the components of O_1 and O_2, while R_1 and R_2, the remaining parts of O_1 and O_2, have measures less than ϵ. (Note carefully that the finiteness of $\mu(O_1)$ and $\mu(O_2)$ is needed at this point—think out carefully *why* O_1 and O_2 can be broken up in the indicated manner.) Then, using some of the earlier results, we obtain: (Convince yourself of the correctness of each step!!) $\mu(O_1 \cup O_2) + \mu(O_1 \cap O_2) - \mu(O_1) - \mu(O_2) = \mu((A_1 \cup A_2) \cup (R_1 \cup R_2)) + \mu((A_1 \cap A_2) \cup (A_1 \cap R_2) \cup (A_2 \cap R_1) \cup (R_1 \cap R_2)) - \mu(A_1 \cup R_1) - \mu(A_2 \cup R_2) \leqslant \mu(A_1 \cup A_2) + \mu(R_1 \cup R_2) + \mu(A_1 \cap A_2) + \mu(A_1 \cap R_2) + \mu(A_2 \cap R_1) + \mu(R_1 \cap R_2) - \mu(A_1) - \mu(A_2) \leqslant \{\mu(A_1 \cup A_2) + \mu(A_1 \cap A_2) - \mu(A_1) - \mu(A_2)\} + \mu(R_1) + \mu(R_2) + \mu(R_2) + \mu(R_1) + \mu(R_1) < \{\mu(A_1 \cup A_2) + \mu(A_1 \cap A_2) - \mu(A_1) - \mu(A_2)\} + 5\epsilon$. Now, by the previous parts of this proof, $\mu(A_1 \cup A_2) + \mu(A_1 \cap A_2) - \mu(A_1) - \mu(A_2) = 0$. Hence, $\mu(O_1 \cup O_2) + \mu(O_1 \cap O_2) - \mu(O_1) - \mu(O_2) < 5\epsilon$. The left side of this inequality has nothing to do with ϵ, while the right side can be pushed as close as desired to zero. Hence,

$$\mu(O_1 \cup O_2) + \mu(O_1 \cap O_2) - \mu(O_1) - \mu(O_2) \leqslant 0. \tag{2-9}$$

On the other hand, using the same breakup as before, we obtain:

$$\begin{aligned}
&\mu(O_1 \cup O_2) + \mu(O_1 \cap O_2) - \mu(O_1) - \mu(O_2) \\
&\qquad \geqslant \mu(A_1 \cup A_2) + \mu(A_1 \cap A_2) - \mu(A_1) - \mu(R_1) - \mu(A_2) - \mu(R_2) \\
&\qquad \geqslant \{\mu(A_1 \cup A_2) + \mu(A_1 \cap A_2) - \mu(A_1) - \mu(A_2)\} - \epsilon - \epsilon \\
&\qquad = -2\epsilon \text{ (since } \{\quad\} = 0).
\end{aligned}$$

Hence, arguing as before about ϵ, we obtain

$$\mu(O_1 \cup O_2) + \mu(O_1 \cap O_2) - \mu(O_1) - \mu(O_2) \geqslant 0. \tag{2-10}$$

Putting (2–9) and (2–10) together, we obtain the desired result.

Remark: The details may appear bewildering, but the *idea* is extremely simple—namely, that any set O of finite measure is almost a union of a finite number of components. Once we have proven the theorem for open sets of this type, we merely have to show that the extra bits R_1 and R_2 cannot destroy the equality.

Exercise

1. Let us define an open interval (a, b) in Q (*not* in R) as the set of all rational numbers satisfying the inequalities $a < x < b$, where a and b are any real (not necessarily rational) numbers; also, let $\mu((a, b))$ be defined as $b - a$. Prove that the interval (a, b) can be expressed as the union of countably many intervals (a_1, b_1), (a_2, b_2), ... such that $\sum_{k=1}^{\infty} \mu((a_k, b_k)) < \mu((a, b))$; in fact, given any positive number ϵ, it is possible to choose the intervals (a_k, b_k) in such a manner that $\sum_{k=1}^{\infty} \mu((a_k, b_k)) < \epsilon$. Note carefully that this result does not contradict Theorem 3; it simply shows that Theorem 3, which is valid in R, is *not* valid in Q. (Hint: Exploit the fact that Q is countably infinite.)

§3. MEASURE OF MORE GENERAL SETS

In this section we shall assign a measure to a class of sets (which will be termed "measurable") which contains all open sets; it will be necessary, of course, to demonstrate that the extended definition of measure agrees with the definition given in §2 when the extended definition is applied to an open set.

Definition 1: *A set S is said to be measurable if for every positive number ϵ there exist sets* O and $\tilde{O}$ such that $S \subseteq O$, $O - S \subseteq \tilde{O}$, and $\mu(\tilde{O}) < \epsilon$. (Note carefully that $\mu(O) = +\infty$ is allowed.) Roughly speaking, this says that S can be covered by an open set without much waste.*

Definition 2: *If S is measurable, $\mu_L(S)$ is defined as* inf $\mu(O)$ *for all sets O which contain S. The number $\mu_L(S)$ (which may be $+\infty$) is called the Lebesgue measure of S.*

Remarks: (*a*) We use the symbol μ_L rather than μ because we have not yet proven that open sets are measurable. Of course, this will be proven very soon, as will the *consistency* of μ and μ_L—i.e., that μ and μ_L coincide for every open set.

(*b*) In some developments of the Lebesgue theory the quantity $\mu_L(S)$ is associated with *every* set S, whether or not S is measurable. (Note that the *definition* of $\mu_L(S)$ is applicable to any set S.) However, most of the following theorems are false if μ_L is not restricted to measurable sets. When μ_L is defined unrestrictedly, it is termed the *outer*

* For emphasis we recall our convention according to which O and $\tilde{O}$ are understood to be open sets.

measure. One then also defines for each set an *inner measure*, and the measurable sets (as previously defined) are those for which the inner and outer measures coincide. (Actually, this statement is true only when the outer measure is finite; a set S can have inner and outer measures both equal to $+\infty$ and yet be unmeasurable. Since we shall confine attention entirely to measurable sets, we shall have no need to consider outer and inner measure, and so we drop the discussion at this point.)

Theorem 1: *Any finite or countably infinite set S is measurable, and its Lebesgue measure is zero.*

PROOF: Given any positive number ϵ, enumerate the points of S, cover the first point with an open interval of length less than $\epsilon/2$, the second point with an open interval of length less than $\epsilon/2^2$, and so forth. The union of these intervals is an open set A, and $\mu(A) < \epsilon/2 + \epsilon/2^2 + \cdots = \epsilon$ (by Theorem 2–3). Letting $O = \tilde{O} = A$ and referring to the two preceding definitions, we see, first, that S is measurable, and, second, that $\mu_L(S) < \epsilon$; since ϵ is arbitrarily small, $\mu_L(S) = 0$.

Theorem 2: *If S and T are both measurable and if $S \subseteq T$, then $\mu_L(S) \leqslant \mu_L(T)$.*

PROOF: Obvious from the definition of μ_L.

Theorem 3: *If S is open it is measurable, and $\mu_L(S) = \mu(S)$.*

PROOF: Referring to the definition of measurability, choose $O = S$ and $\tilde{O} = \emptyset$. This shows that S is measurable. Now, referring to the definition of μ_L (of a measurable set), we observe that S is one of the open sets that contain S, and so $\inf_{S \subseteq O} \mu(O)$ cannot *exceed* $\mu(S)$. Hence, $\mu_L(S) \leqslant \mu(S)$. Now, if $\mu_L(S)$ were less than $\mu(S)$, this would imply the existence of an *open* set T containing S such that $\mu(T) < \mu(S)$, and this would contradict Theorem 2–1. Hence, $\mu_L(S) = \mu(S)$. (Thus, as promised previously, we have shown that μ_L and μ are consistently defined for open sets. We may, therefore, henceforth denote the measure of *any* measurable set S by $\mu(S)$, omitting the subscript L.)

Theorem 4: *If $S_1, S_2, \ldots$ are each measurable, so is $S_1 \cup S_2 \cup \cdots$. Furthermore, $\mu(S_1 \cup S_2 \cup \cdots) \leqslant \mu(S_1) + \mu(S_2) + \cdots$.*

PROOF: (*a*) Given $\epsilon > 0$, associate with each set S_k a set O_k and a set $\tilde{O}_k$ such that $S_k \subseteq O_k$, $O_k - S_k \subseteq \tilde{O}_k$, $\mu(\tilde{O}_k) < \epsilon/2^k$. Then, clearly, $\bigcup S_k \subseteq \bigcup O_k$, $\bigcup O_k - \bigcup S_k \subseteq \bigcup \tilde{O}_k$, and $\mu(\bigcup \tilde{O}_k) < \epsilon$. Hence, $\bigcup S_k$ is measurable.

(*b*) Given $\epsilon > 0$, choose O_k (not necessarily the same O_k as in part (*a*)) such that $\mu(O_k) \leqslant \mu(S_k) + \epsilon/2^k$ and $S_k \subseteq O_k$. Then, by Theorem 2–3, $\mu(\bigcup O_k) \leqslant \sum \mu(O_k) \leqslant \epsilon + \sum \mu(S_k)$. Since $\bigcup O_k$ is an open set containing $\bigcup S_k$, it follows from the definition of measure that $\mu(\bigcup S_k) \leqslant \mu(\bigcup O_k)$. Hence $\mu(\bigcup S_k) \leqslant \epsilon + \sum \mu(S_k)$. Since ϵ can be chosen as small as we wish, we obtain $\mu(\bigcup S_k) \leqslant \sum \mu(S_k)$.

Theorem 5: *If O is open and $\epsilon > 0$ is given, there exists a closed set F such that $F \subseteq O$ and $\mu(O - F) < \epsilon$. (Note that $O - F$ is open, and hence measurable.)*

PROOF: (*a*) First suppose that $\mu(O) < +\infty$ and that O consists of a finite number of components, say $O = (a_1, b_1) \cup (a_2, b_2) \cup \cdots \cup (a_n, b_n)$. Choose compact subintervals $[a_1', b_1'], [a_2', b_2'], \ldots, [a_n', b_n']$ such that $\mu((a_1, b_1) - [a_1', b_1']) < \epsilon/2$, $\mu((a_2, b_2) - [a_2', b_2']) < \epsilon/2^2$, and so forth, and let F be the union of these compact subintervals. Then F is closed and $O - F$ consists of $2n$ open intervals, $(a_1, a_1'), (b_1', b_1), (a_2, a_2'), (b_2', b_2), \ldots$, and their total length is less than $\epsilon/2 + \epsilon/2^2 + \cdots + \epsilon/2^n < \epsilon$. Hence, $\mu(O - F) < \epsilon$.

(*b*) Now we still suppose that $\mu(O) < +\infty$ but we allow O to have infinitely many components. We write O in the form $A \cup B$, where A consists of a *finite* number of the components of O and $\mu(B) < \epsilon/2$. Then by part (*a*) we can construct a closed subset F of A such that $\mu(A - F) < \epsilon/2$. Thus, $\mu(O - F) = \mu(B \cup (A - F)) = \mu(B) + \mu(A - F) < \epsilon/2 + \epsilon/2 = \epsilon$.

(*c*) Now suppose that $\mu(O) = +\infty$ (so that O is certainly unbounded in at least one direction). Let $O_n = O \cap (n, n + 1)$, $n = 0, \pm 1, \pm 2, \ldots$. For each n construct a closed subset F_n of O_n such that $\mu(O_n - F_n) < 2^{-|n|}\epsilon/3$; this is certainly possible, by parts (*a*) and (*b*). Now, even though the union of infinitely many closed sets is *not* always closed, nevertheless $\bigcup_{n=-\infty}^{\infty} F_n$ *is* closed. (Why?) Observe that $O - F = \{\bigcup_{n=-\infty}^{\infty} (O_n - F_n)\} \cup$ (set of integers belonging to O), and so $\mu(O - F) \leqslant \{\sum_{n=-\infty}^{\infty} \mu(O_n - F_n)\} + \mu(\text{set of all integers}) = \sum_{n=-\infty}^{\infty} \mu(O_n - F_n) < \sum_{n=-\infty}^{\infty} 2^{-|n|}\epsilon/3 = \epsilon$, and so the proof is complete.

Theorem 6: *If A is measurable and $\epsilon > 0$ is given, there exists a closed subset F of A such that $A - F$ can be covered by an open set of measure $<\epsilon$.*

PROOF: Referring to the definition of measurability, form open sets O and $\tilde{O}$ such that $A \subseteq O$, $O - A \subseteq \tilde{O}$, $\mu(\tilde{O}) < \epsilon/2$. Next, construct a closed set G such that $G \subseteq O$ and $\mu(O - G) < \epsilon/2$. Let $F = G - \tilde{O}$. Then F is closed, $F \subseteq A$, and $A - F \subseteq \tilde{O} \cup (O - G)$. (These two inclusions should be checked carefully.) Clearly, $\tilde{O} \cup (O - G)$ is open, and its measure is $\leqslant \mu(\tilde{O}) + \mu(O - G)$, which in turn is less than ϵ. Thus, the proof is complete.

Remark: Roughly speaking, Theorem 5 asserts that any *open* set can be almost exhausted by a closed set, while Theorem 6 asserts that any *measurable* set can be almost exhausted by a closed set.

Theorem 7: *A is measurable iff A^c is measurable. (In particular, since all open sets are measurable, it now follows that all closed sets are measurable.)*

PROOF: Suppose A is measurable. Given $\epsilon > 0$, there exists a closed set F such that $F \subseteq A$ and $A - F$ can be covered by an open set U of measure $<\epsilon$. Then $F^c \supseteq A^c$ and $F^c - A^c = A - F \subseteq U$. Thus, A^c has been covered by an open set, F^c, with $F^c - A^c$ contained in an open set, U, of measure $<\epsilon$. Hence A^c is measurable.

Conversely, if A^c is measurable, we now know that $(A^c)^c$ is measurable; but $(A^c)^c = A$.

Theorem 8: *If $A_1, A_2, \ldots$ are each measurable, so is $\bigcap A_k$.*

PROOF: $A_1^c, A_2^c, \ldots$ are each measurable, by Theorem 7, and, by Theorem 4, $A_1^c \cup A_2^c \cup A_3^c \cup \cdots$ is measurable. By Theorem 7,

$$(A_2^c \cup A_2^c \cup \cdots)^c$$

is also measurable—but this is $\bigcap A_k$.

Theorem 9: *If A and B are measurable, so is $A - B$.*

PROOF: B^c is measurable by Theorem 7, so $A \cap B^c$ is measurable by Theorem 8; but $A \cap B^c = A - B$.

Theorem 10: *If A and B are measurable and disjoint, $\mu(A \cup B) = \mu(A) + \mu(B)$. (By Theorem 4 we know that $A \cup B$ is measurable.)*

PROOF: If either $\mu(A)$ or $\mu(B)$ is $+\infty$ there is nothing to prove, so we assume that $\mu(A)$ and $\mu(B)$ are both finite. Given $\epsilon > 0$, find sets O_1 and $\tilde{O}_1$ such that $A \subseteq O_1$, $O_1 - A \subseteq \tilde{O}_1$, $\mu(\tilde{O}_1) < \epsilon$. Since $O_1 - A$ is certainly measurable (by Theorem 9), $\mu(O_1 - A) \leqslant \mu(\tilde{O}_1) < \epsilon$. Call this set (i.e., $O_1 - A$) R_1. Thus, $O_1 = A \cup R_1$, where $\mu(R_1) < \epsilon$, and so $\mu(A) \leqslant \mu(O_1) \leqslant \mu(A) + \mu(R_1) < \mu(A) + \epsilon$. Similarly, we can find a set O_2 such that $B \subseteq O_2$ and $\mu(O_2 - B) < \epsilon$, and so $\mu(B) \leqslant \mu(O_2) < \mu(B) + \epsilon$. But, by Theorem 2–4, $\mu(O_1 \cup O_2) + \mu(O_1 \cap O_2) - \mu(O_1) - \mu(O_2) = 0$. On the other hand, $\mu(O_1 \cup O_2) + \mu(O_1 \cap O_2) - \mu(O_1) - \mu(O_2) = \mu((A \cup B) \cup (R_1 \cup R_2)) + \mu((A \cap B) \cup (A \cap R_2) \cup (B \cap R_1) \cup (R_1 \cap R_2)) - \mu(O_1) - \mu(O_2) < \mu(A \cup B) + \mu(R_1) + \mu(R_2) + \mu(R_2) + \mu(R_1) + \mu(R_1) - \mu(A) - \mu(B) < 5\epsilon + \{\mu(A \cup B) - \mu(A) - \mu(B)\}$, and so $\mu(A \cup B) - \mu(A) - \mu(B) > -5\epsilon$. Since ϵ can be chosen arbitrarily close to zero, we obtain $\mu(A \cup B) - \mu(A) - \mu(B) \geqslant 0$, or $\mu(A) + \mu(B) \leqslant \mu(A \cup B)$. On the other hand, we can find open sets U_1 and U_2 such that $A \subseteq U_1$, $B \subseteq U_2$, and $\mu(U_1) < \mu(A) + \epsilon$, $\mu(U_2) < \mu(B) + \epsilon$. Since $U_1 \cup U_2$ is an open set containing $A \cup B$, $\mu(A \cup B) \leqslant \mu(U_1 \cup U_2) \leqslant \mu(U_1) + \mu(U_2) < \mu(A) + \epsilon + \mu(B) + \epsilon = \mu(A) + \mu(B) + 2\epsilon$. Since ϵ is arbitrarily small, $\mu(A \cup B) \leqslant \mu(A) + \mu(B)$. Combining this with an earlier result obtained during this proof, we obtain $\mu(A \cup B) = \mu(A) + \mu(B)$.

Remark: Note where the disjointness of A and B is used. As in Theorem 2–4, the details may appear bewildering, but the essential idea is very simple.

Theorem 11: *If $A_1, A_2, \ldots$ are each measurable and pairwise disjoint, then $\mu(A_1 \cup A_2 \cup \cdots) = \mu(A_1) + \mu(A_2) + \cdots$. (The measurability of $A_1 \cup A_2 \cup \cdots$ is known from Theorem* 4.)

PROOF: By Theorem 4 we already know that the left side cannot exceed the right side. It therefore suffices to prove the reverse inequality. From Theorem 10 and finite induction we know that, for $1 < n < +\infty$,

$$\mu(A_1 \cup A_2 \cup \cdots \cup A_n) = \mu(A_1) + \mu(A_2) + \cdots + \mu(A_n).$$

Since $\bigcup_{k=1}^{\infty} A_k \supseteq \bigcup_{k=1}^{n} A_k$, we know that

$$\mu(A_1 \cup A_2 \cup \cdots) \geqslant \mu(A_1) + \mu(A_2) + \cdots + \mu(A_n).$$

(Note that the left side has nothing to do with n.) Letting n increase without bound, we obtain the desired inequality, namely,

$$\mu(A_1 \cup A_2 \cup \cdots) \geqslant \sum_{k=1}^{\infty} \mu(A_k).$$

Theorem 12: *If $B \subseteq A$, if both sets are measurable, and if $\mu(A)$ is finite, then $\mu(A - B) = \mu(A) - \mu(B)$.*

PROOF: By Theorem 9, $A - B$ is measurable. We may write $A = B \cup (A - B)$, so by Theorem 10 we are sure that $\mu(A) = \mu(B) + \mu(A - B)$, and so $\mu(A - B) = \mu(A) - \mu(B)$. (Note carefully that the last step is invalid if $\mu(A) = +\infty$.)

Theorem 13: *If $A_n \subseteq A_{n+1}$ and if each A_n is measurable, then their union is measurable, and $\mu(\bigcup_{n=1}^{\infty} A_n) = \lim_{n\to\infty} \mu(A_n)$.*

PROOF: (*a*) It follows directly from Theorem 4 that the union is measurable.

(*b*) Since $A_n \subseteq A_{n+1}$, $\mu(A_n) \leqslant \mu(A_{n+1})$, and so $\lim_{n\to\infty} \mu(A_n)$ must exist, either as a finite number or $+\infty$.

(*c*) If any particular A_n has infinite measure, say A_k, then clearly $\mu(A_m) = +\infty$ whenever $m \geqslant k$, and then $\mu(\bigcup_{j=1}^{\infty} A_j)$ must clearly be $+\infty$, which is also $\lim_{n\to\infty} \mu(A_n)$.

(*d*) Therefore, we may assume that each A_n has finite measure. Then by Theorem 11 we obtain

$$\mu\left(\bigcup_{n=1}^{\infty} A_n\right) = \mu(A_1) + \mu(A_2 - A_1) + \mu(A_3 - A_2) + \cdots$$

Taking any partial sum of the infinite series on the right side and using Theorem 12, we obtain $\mu(A_1) + \mu(A_2 - A_1) + \cdots + \mu(A_n - A_{n-1}) = \mu(A_1) + \mu(A_2) - \mu(A_1) + \mu(A_3) - \mu(A_2) + \cdots + \mu(A_n) - \mu(A_{n-1}) = \mu(A_n)$. Thus, the sum of the infinite series must be $\lim_{n\to\infty} \mu(A_n)$, and so we have the desired result.

Theorem 14: *If $A_n \supseteq A_{n+1}$, if all the A's are measurable, and if $\mu(A_1)$ is finite, then $\mu(A_1 \cap A_2 \cap \cdots) = \lim_{n\to\infty} \mu(A_n)$.*

PROOF: (*a*) Since the sets are nested and $\mu(A_1)$ is finite, each A_n has finite measure and these measures form a monotone non-increasing sequence of real non-negative numbers, and so the limit of $\mu(A_n)$ exists. Also, by Theorem 8, $A_1 \cap A_2 \cap \cdots$ is measurable. Now, observe that A_1 can be expressed as a *disjoint* union in the following way:

$$A_1 = \left(\bigcap_{k=1}^{\infty} A_k\right) \cup (A_1 - A_2) \cup (A_2 - A_3) \cup \cdots.$$

By Theorem 11, we obtain

$$\mu(A_1) = \mu\left(\bigcap_{k=1}^{\infty} A_k\right) + \sum_{k=1}^{\infty} \mu(A_k - A_{k+1})$$

$$= \mu\left(\bigcap_{k=1}^{\infty} A_k\right) + \lim_{n\to\infty} \sum_{k=1}^{n} \mu(A_k - A_{k+1}),$$

and by Theorem 12 we then obtain

$$\mu(A_1) = \mu\left(\bigcap_{k=1}^{\infty} A_k\right) + \lim_{n\to\infty} \{\mu(A_1) - \mu(A_2) + \mu(A_2) - \mu(A_3) + \cdots + \mu(A_n) - \mu(A_{n+1})\}$$
$$= \mu\left(\bigcap_{k=1}^{\infty} A_k\right) + \mu(A_1) - \lim_{n\to\infty} \mu(A_n).$$

Cancelling out $\mu(A_1)$ from both sides, we obtain

$$\mu\left(\bigcap_{k=1}^{\infty} A_k\right) = \lim_{n\to\infty} \mu(A_n).$$

Remark: Note that the finiteness of $\mu(A_1)$ was needed in order to use Theorem 12. However, one might ask whether some other argument might give the desired result *without* assuming the finiteness of $\mu(A_1)$. The answer is *NO*! (Why?) (Actually, the theorem clearly holds if the first million A's have infinite measure, provided that for *some* index k the set A_k has finite measure. Of course, all the following A's will then also have finite measure.)

Definition 2: *A null-set (not to be confused with the empty set) is a set of measure zero. (From the definition of measurability and from the definition of measure it is easily seen that A is a null-set iff for every $\epsilon > 0$ there exists an open set O of measure $<\epsilon$ such that $A \subseteq O$. Clearly, any subset of a null-set is both measurable and a null-set. Also, by Theorem 4, any finite or countably infinite union of null-sets is a null-set.)*

Definition 3: *If a particular set A is under consideration and if something happens at all points of A with the exception of a null-set, we say that this happens almost everywhere on A. For example, if a sequence of functions diverges on a certain null-set $B \subseteq A$ and converges everywhere in $A - B$, we say that the sequence converges almost everywhere in A or at almost all points of A. The abbreviation* a.e. *is often used, and occasionally the French abbreviation* p.p. *(presque partout) is employed.*

Definition 4: *Two sets E and F are said to be equivalent if $E \,\Delta F$ is a null-set, where $E \,\Delta F = (E - F) \cup (F - E)$. Of course, this use of the word equivalent is entirely different from its use in set theory, where it means that a one-to-one correspondence exists between two given sets.*

Remark: If E and F are equivalent, then F can be obtained by removing a null-set from E and then adding a null-set to the resulting set. Of course, E and F can be interchanged in this statement. It therefore follows without difficulty that this is indeed an equivalence relation. Note that we do *not* assume that E and F are measurable in this definition.

Theorem 15: *If E and F are equivalent and if either one is measurable, so is the other, and their measures are equal.*

PROOF: Since $E - F$ and $F - E$ are subsets of $E \Delta F$, it follows from the definition that they are both null-sets. Suppose E is measurable; then $E \cup (F - E)$ is measurable; but $E \cup (F - E) = E \cup F$, and so $E \cup F$ is measurable. But $F = (E \cup F) - (E - F)$, so F is measurable by Theorem 9. Now, $\mu(E \cup F) = \mu(E) + \mu(F - E) = \mu(E)$, while $\mu(F) = \mu(E \cup F) - \mu(E - F) = \mu(E \cup F)$. Hence, $\mu(E) = \mu(F)$. (Note that this argument works even when $\mu(E \cup F) = +\infty$.)

Exercises

1. (*a*) Show that a set A is a null-set iff for every positive number ϵ it is possible to find a countably infinite collection $\{O_1, O_2, O_3, \ldots\}$ of open intervals such that $\sum_{k=1}^{\infty} \mu(O_k) < \epsilon$ and each point of A belongs to at least one of the O's.

   (*b*) Show that the preceding result remains correct if "at least one" is replaced by "infinitely many."

   Note that this exercise provides a definition of a null-set which does not presuppose a knowledge of the concept of measure, or even of a measurable set.

2. (*a*) Prove that measurability and measure are both *translation-invariant*; that is, if A is measurable and if B is obtained by adding to each member of A a fixed number, then B is measurable and $\mu(B) = \mu(A)$.

   (*b*) Suppose that a non-negative number $\nu(A)$ is associated with each measurable set, that $\nu(A)$ is translation-invariant, and that it possesses the countable additivity property (Theorem 11). Prove that $\nu(A) = c\mu(A)$ for some fixed non-negative number c.

3. (*a*) Let $A = [0, 1)$ and let two members of A be called equivalent if their difference is rational. Prove that this definition determines an equivalence relation on A.

   (*b*) Let A be decomposed into the equivalence classes determined by the equivalence relation defined in (*a*), and let one member be chosen from each equivalence class. Show that the set S formed in this manner is not measurable. Hint: Suppose that S is measurable. For each rational number r in A (including zero) let S_r be the set obtained by adding r to each member of S. By using a slight extension of part (*a*) of the preceding exercise, show that $\bigcup S_r$ must have measure equal to one, but that this leads to a contradiction of the assumption $\mu(S) = 0$ and also of the assumption $\mu(S) > 0$.

Thus, the existence of an unmeasurable set has been demonstrated.

4. A G_δ is a subset of R which can be expressed as the intersection of a countably infinite collection of open sets, while an F_σ is a subset of R which can be expressed as the union of a countably infinite collection of closed sets. Show that every measurable set can be expressed as the difference of a G_δ and a null-set and also as the disjoint union of an F_σ and a null-set.

§4. MEASURABLE FUNCTIONS

In the two preceding sections we have developed the essential parts of the theory of measure. In this section we develop the theory of measurable functions, which forms the link between measure and integration.

Definition 1: *A real-valued function f is said to be measurable if for every real number α the set $\{x \mid f(x) < \alpha\}$ (which can also be written $f^{-1}((-\infty, \alpha))$ is measurable.*

Theorem 1: *If f is measurable, its domain E is measurable.*

PROOF: Clearly, $E = \bigcup_{n=1}^{\infty} f^{-1}((-\infty, n))$; thus E is a union of countably many measurable sets, and is therefore measurable, by Theorem 3–4.

Theorem 2: *If f satisfies any one of the following four conditions, it satisfies the other three, and so any one of the four may be used in the definition of measurability:*

(*a*) *$f^{-1}((-\infty, \alpha))$ is measurable for every real number α;*

(*b*) *$f^{-1}((-\infty, \alpha])$ is measurable for every real number α;*

(*c*) *$f^{-1}((\alpha, +\infty))$ is measurable for every real number α;*

(*d*) *$f^{-1}([\alpha, +\infty))$ is measurable for every real number α.*

PROOF: Suppose (*a*) is true. Clearly $f^{-1}((-\infty, \alpha])$ can be expressed as $\bigcap_{n=1}^{\infty} f^{-1}((-\infty, \alpha + 1/n))$, so by Theorem 3–8 it follows that $f^{-1}((-\infty, \alpha])$ is measurable. Next (still assuming that (*a*) is true), we see that $f^{-1}((\alpha, +\infty)) = E - f^{-1}((-\infty, \alpha])$, and so $f^{-1}((\alpha, +\infty))$ is the difference of two measurable sets. Next, we observe that $f^{-1}([\alpha, +\infty)) = E - f^{-1}((-\infty, \alpha))$. Thus, $f^{-1}([\alpha, +\infty))$ is measurable if (*a*) is true.

Hence, (a) guarantees (b), (c), and (d). Similarly, each of (b), (c), or (d) guarantees the three remaining conditions.

Definition 2: *Two functions f and g are said to be equivalent if their domains are equivalent and if they agree almost everywhere (on the intersection of their respective domains).*

Theorem 3: *If f and g are equivalent and if one of them is measurable, so is the other.*

PROOF: Trivial.

Theorem 4: *If f and g are both measurable (and are defined on the same domain) and if c is any real number, then the following functions are also measurable:*

(a) cf, (b) $f + g$, (c) $f - g$, (d) f^2, (e) fg.

PROOF: (a) Trivial if $c = 0$. If $c > 0$,

$$\{x \mid cf(x) < \alpha\} = \{x \mid f(x) < \alpha/c\},$$

and so the left side is measurable. If $c < 0$,

$$\{x \mid cf(x) < \alpha\} = \{x \mid f(x) > \alpha/c\},$$

and so, again, the left side is measurable.

(b) Suppose that, for some x, $f(x) + g(x) < \alpha$. Then it is possible to find *rational* numbers s and t such that $f(x) < s$, $g(x) < t$, and $s + t < \alpha$. (This is so because the rationals are dense in R.) Conversely, if $f(x) < s$ and $g(x) < t$ and $s + t < \alpha$, then, clearly, $f(x) + g(x) < \alpha$. Thus,

$$\{x \mid f(x) + g(x) < \alpha\} = \bigcup_{\substack{r,s \text{ rational} \\ r+s<\alpha}} \{x \mid f(x) < r\} \cap \{x \mid g(x) < s\}.$$

Since f and g are measurable, each $\{\ \}$ appearing in the preceding union is measurable, hence each set $\{\ \} \cap \{\ \}$ is measurable, and finally the entire right side, being a union of *countably many* measurable sets, is measurable. Hence, $f + g$ is measurable.

(c) By (a), $-g$ is measurable, and by (b) we see that $f + (-g)$ is measurable.

(d) For $\alpha \leqslant 0$, the set $\{x \mid f^2(x) < \alpha\}$ is empty, hence measurable. For $\alpha > 0$, the set $\{x \mid f^2(x) < \alpha\}$ can be written as

$$\{x \mid f(x) < \sqrt{\alpha}\} \cap \{x \mid f(x) > -\sqrt{\alpha}\}.$$

Since each { } is measurable (see Theorem 2), their intersection is measurable, and so f^2 is measurable.

(*e*) Note that $fg = \frac{1}{4}\{(f+g)^2 - (f-g)^2\}$ and observe, with the aid of (*a*), (*b*), (*c*), and (*d*), that the right side is measurable.

Theorem 5: *Let* $\{f_n\}$ *be a finite or countably infinite collection of measurable functions, all defined on a fixed domain. If this collection is infinite, we assume that the set of numbers* $\{f_n(x)\}$ *is bounded for each* x *(but not necessarily uniformly bounded on the entire domain). Then the functions* $\sup \{f_n\}$ *and* $\inf \{f_n\}$ *are also measurable.*

PROOF: It obviously suffices to discuss $\sup \{f_n\}$. Call this function f. Then $\{x \mid f(x) > \alpha\} = \{x \mid f_1(x) > \alpha\} \cup \{x \mid f_2(x) > \alpha\} \cup \cdots$. Each of the sets { } on the right is measurable, and, since only a finite or countably infinite number of sets are involved, their union is measurable. Hence, f is measurable.

Corollary: *If* f *is measurable, so is* $|f|$.

PROOF: Observe that $|f| = \sup \{f, -f\}$ and use Theorems 4 and 5.

The concept of limit superior (lim sup, $\overline{\lim}$) and limit inferior (lim inf, $\underline{\lim}$) should be known to the reader; a brief presentation of this topic is given in Appendix *G*, and it should be studied carefully at this time if the reader is not thoroughly acquainted with it.

Theorem 6: *Let* $\{f_n\}$ *be a sequence of measurable functions (all defined on the same domain) and suppose that for each* x *the sequence* $\{f_n(x)\}$ *is bounded (above and below). Then the functions* $\overline{\lim}\, f_n$ *and* $\underline{\lim}\, f_n$ *are also measurable.*

PROOF: Let $g_1 = \sup \{f_1, f_2, \ldots\}$, $g_2 = \sup \{f_2, f_3, \ldots\}$, and so forth. By Theorem 5, $g_1, g_2, \ldots$ are all measurable, and, *again* by Theorem 5, we see that $\inf g_k$ is measurable. Now (cf. Appendix *G*), $\inf g_k = \overline{\lim}\, f_n$. Thus $\overline{\lim}\, f_n$ is measurable, and, similarly, so is $\underline{\lim}\, f_n$.

Corollary: *If the functions* $f_1, f_2, \ldots$ *are all measurable and converge (to a finite limit) everywhere (on the common domain of all the* f_k*'s), the limit function is also measurable.*

PROOF: If the given sequence converges everywhere, then the limit coincides both with the $\overline{\lim}$ and the $\underline{\lim}$, each of which is measurable by Theorem 6.

Theorem 7: *If f is measurable and if A is a measurable subset of the domain of f, then $f|A$ (the restriction of f to A) is also measurable.*

PROOF: Obvious.

We now indicate briefly how the foregoing development must be modified in order to take account of functions which assume infinite values. Such a modification is necessary, in particular, in order to be able to assign an upper limit and a lower limit to every sequence of real numbers and to every sequence of real-valued functions.

Definition 3: *The extended real number system, denoted R^*, consists of R together with the new symbols, $+\infty$ and $-\infty$, with the obvious ordering* (i.e., *for any real number a, $-\infty < a < +\infty$*) *and with the obvious (or almost obvious) arithmetic: For any real number a, $a + (\pm\infty) = \pm\infty$; $a \cdot (\pm\infty) = \pm\infty$ if $a > 0$, $a \cdot (\pm\infty) = \mp\infty$ if $a < 0$, $a \cdot (\pm\infty) = 0$ if $a = 0$; $(+\infty) \cdot (+\infty) = +\infty$, $(+\infty) \cdot (-\infty) = -\infty$; $(+\infty) + (-\infty)$ is not defined. The topology (i.e., the collection of open sets) assigned to R^* is a rather obvious extension of the topology of R* (*which is determined by the metric which we have defined on R in Chapter* 1). *All sets of the form $(a, +\infty]$ and all sets of the form $[-\infty, a)$ are declared to be open (for every real number a), and a subset A of R^* is declared to be open iff A can be expressed as a union (not necessarily countable) $\bigcup B_\alpha$, where each B is a finite intersection of the particular open sets described at the beginning of this sentence.*† (*Note carefully that every open subset of R is also an open subset of R^*, but this is not true for closed subsets; for example, R is a closed subset of R, but not of R^*.*)

Definition 4: *A function defined on R, or on a subset of R, with values in R^*, is called measurable if the inverse image of every set of the form $(a, +\infty]$ and the inverse image of every set of the form $[-\infty, a)$ are measurable (for every real number a). (Note carefully that the domain of the function is still restricted to R; only the allowed range has been enlarged. Also, there should be no difficulty in seeing that when the function does not assume the values $+\infty$ or $-\infty$ the two definitions of measurability are consistent.)*

Remark: All the previous results about measurable functions (Theorems 2 to 7) continue to remain valid, except that in parts (b) and (c) of Theorem 4 we must restrict the domains suitably, to eliminate points at which we have $(\pm\infty) - (\pm\infty)$ or $(\pm\infty) + (\mp\infty)$. We shall see that this is a small price to pay for the advantages gained by working in R^*.

† The reader acquainted with the elements of topology will recognize that the collection of all sets of the form $(a, +\infty]$ or $[-\infty, a)$ constitutes an *open subbase* of the topology of R^*.

Exercises

1. Given any subset E of R, χ_E, the *characteristic function of* E, is the function defined as follows: $\chi_E(x) = 1$ if $x \in E$, $\chi_E(x) = 0$ if $x \notin E$. Show that a set E is measurable iff χ_E is measurable.

2. Let E be any measurable subset of a finite interval A. Given any positive number ϵ, show that there exists a subset $\tilde{E}$ of A consisting of a finite collection of intervals such that $\mu(E \,\Delta \tilde{E}) < \epsilon$.

3. Let E be any measurable subset of a finite interval A. Given any positive number ϵ, show that there exists a continuous function f which satisfies the conditions $0 \leqslant f(x) \leqslant 1$ everywhere on A and which coincides with χ_E except on a set of measure less than ϵ. (Hint: Use part (*b*) of Exercise 9–5 of Chapter 1.)

4. A *simple function* is one whose range consists of a finite set of real numbers. A *step-function* is a function which can be expressed as a finite linear combination of characteristic functions of intervals. (Note that a step-function must be simple.) Show that if f is any measurable simple function defined on a finite interval A and if any positive number ϵ is given, there exist a step-function g and a continuous function h such that g and h both agree with f everywhere on A except for a subset of measure less than ϵ. Furthermore, one may impose on g and h the restriction that their ranges should lie in the interval $[\min f, \max f]$.

5. Let the sequence of measurable functions $f_1, f_2, f_3, \ldots$, each defined on a set A of finite measure, converge everywhere in A to the function f. (All functions are assumed to take values in R, not in R^*.) Given any positive numbers δ and ϵ, show that there exists an index N such that the inequality

$$|f(x) - f_N(x)| < \epsilon$$

holds everywhere in A except on a subset of measure less than δ.

6. (Egoroff's Theorem): Prove the following strengthened version of the preceding exercise: Under the same hypotheses as in Exercise 5, for any positive number δ there exists a subset B of A such that $\mu(B) < \delta$ and the given sequence of functions converges *uniformly* on $A - B$.

7. Prove that the condition that $\mu(A)$ be finite cannot be omitted in either of the two preceding exercises.

8. (Lusin's Theorem): Let f be bounded and measurable on a finite interval A and let a positive number δ be given. Show that there exists a continuous function g defined on A such that f and g agree everywhere except on a set of measure less than δ. Furthermore, one may impose on g the condition that its range should be contained in the interval $[\inf f, \sup f]$. Hint: Approximate f by a sequence of simple functions, then approximate the simple functions by continuous functions, and finally invoke Egoroff's theorem. (Actually, the conditions that f be bounded and that $\mu(A)$ be finite can be omitted.)

Remark: Egoroff's theorem can be stated, imprecisely but vividly, in the form: Every convergent sequence of measurable functions is almost uniformly convergent. Similarly, Lusin's theorem asserts that every measurable function is almost continuous, while Definition 3–1 characterizes measurable sets as those which are almost open. These three approximate truths constitute the "Three Principles" of Littlewood, who emphasizes that they play a major role in the development of the theory of functions of a real variable.

§5. INTEGRATION OF NON-NEGATIVE FUNCTIONS

We are now in position to undertake the development of the Lebesgue integral. Always keep in mind that, since we allow sets of infinite measure, expressions of the form $(a) \cdot (+\infty)$ may be encountered, where a is a (finite) real number. In many treatments of the subject one begins with functions defined on sets of finite measure and later removes this restriction, but we believe it preferable to admit sets of infinite measure from the very beginning. Also, recall that functions may assume values in R^*—i.e., $+\infty$ and $-\infty$ are permitted functional values.

Definition 1: *A simple function is a function whose domain is measurable and whose range consists of a finite number of finite non-negative numbers. (Note that $+\infty$ is not allowed as a value of the function.)*

It should be remarked that in some books the restriction to non-negative values is not imposed. (Actually, the present definition does not agree with that given in Theorem 4–4; the change has been made for convenience, and no confusion should result.)

Theorem 1: *Let f be a non-negative measurable function, defined on a set A (which must be measurable, by Theorem 4–1). There exists a* *sequence* $s_1(x), s_2(x), \ldots$ *of measurable simple functions (all defined on* A*) such that:*

(*a*) $0 \leqslant s_1(x) \leqslant s_2(x) \leqslant \cdots$ *everywhere on* A,

(*b*) $s_n(x) \to f(x)$ *everywhere on* A.

PROOF: For each positive integer n, form the set F_n and the sets $E_{n,1}, E_{n,2}, \ldots, E_{n,n\cdot 2^n}$ as follows:

$$F_n = \{x \mid f(x) \geqslant n\}, \qquad E_{n,k} = \left\{x \,\middle|\, \frac{k-1}{2^n} \leqslant f(x) < \frac{k}{2^n}\right\}.$$

(Clearly, A is the disjoint union of F_n and the sets $E_{n,k}$, and all these sets are measurable.) Now, let

$$s_n(x) = \begin{cases} n & \text{if} \quad x \in F_n, \\ \dfrac{k-1}{2^n} & \text{if} \quad x \in E_{n,k}. \end{cases}$$

The sequence of functions $s_1, s_2, s_3, \ldots$ clearly satisfy conditions (*a*) and (*b*).

Definition 2: *Let* s *be a measurable simple function with domain* A*; let the distinct values of* s *be* $\alpha_1, \alpha_2, \ldots, \alpha_m$*; and let the sets* $s^{-1}(\{\alpha_k\})$ *be denoted* A_k*, so that* A *is the disjoint union of the* A_k*'s and the* A_k*'s are measurable. We define the integral of* s *over* A *as follows:*

$$\int_A s = \sum_{k=1}^{m} \alpha_k \mu(A_k).$$

If B *is any measurable subset of* A*, we define* $\int_B s$ *to be* $\int_B \tilde{s}$*, where* $\tilde{s}$ *is the restriction of* s *to* B*. (Clearly* $\tilde{s}$ *is also simple and measurable, so this definition makes sense.)*

Theorem 2:

$$\int_B s = \sum_{k=1}^{m} \alpha_k \mu(B \cap A_k).$$

PROOF: Trivial.

Theorem 3: *Let* s *be a simple measurable function defined on* A *and let* A *be the disjoint union of the measurable sets* $B_1, B_2, \ldots$ *(where we allow either a finite or countably infinite number of* B*'s). Then*

$$\int_A s = \sum_k \int_{B_k} s.$$

(If there are infinitely many B's the convergence of the sum on the right is part of the theorem.)

PROOF: Trivial.

Theorem 4: *If s_1 and s_2 are measurable simple functions defined on A and if $s_1 \geqslant s_2$ everywhere on A, then $\int_A s_1 \geqslant \int_A s_2$.*

PROOF: (*a*) Obvious from the definition if s_1 and s_2 are both constant functions, or even if one of them is constant.

(*b*) If neither s_1 nor s_2 is constant, split A into disjoint sets $B_1, B_2, \ldots$ (finitely many) on which both s_1 and s_2 are constant. (It is obvious how to do this.) Then, using Theorem 3 and part (*a*) of the present proof, we obtain

$$\int_A s_1 = \sum_k \int_{B_k} s_1 \geqslant \sum_k \int_{B_k} s_2 = \int_A s_2,$$

which is the desired result.

Definition 3: *Let f be any non-negative measurable function defined on A. (Note that we allow $+\infty$ as a value of f.) Then we define*

$$\int_A f = \sup \int_A s,$$

where s ranges over all measurable simple functions satisfying $s \leqslant f$ everywhere on A. If B is a measurable subset of A, we define

$$\int_B f = \sup \int_B s.$$

Of course, $\int_A f$ is called the integral of f over A.

Theorem 5: *The two preceding definitions are consistent; i.e., if f is simple, these two definitions assign the same value to $\int_A f$.*

PROOF: (*a*) If f is simple, it is one of the functions s appearing in the second definition; hence, $\int_A f$ (according to second definition) $\geqslant \int_A f$ (according to first definition).

(*b*) If f is simple and $s \leqslant f$ everywhere on A, then, by Theorem 4, $\int_A f \geqslant \int_A s$. Hence, $\int_A f$ (according to second definition) $\leqslant \int_A f$ (according to first definition).

(*c*) Now we merely combine (*a*) and (*b*).

Theorem 6: *Assuming that all sets involved are measurable and that all functions are measurable and non-negative:*

(*a*) *If* $f \leqslant g$ *everywhere, then* $\int_E f \leqslant \int_E g$.

(*b*) *If* $A \subseteq B$, *then* $\int_A f \leqslant \int_B f$.

(*c*) *If c is a non-negative constant (even* $+\infty$ *allowed),*

$$\int_E cf = c\int_E f.$$

(*d*) If $f \equiv 0$ *on E, then* $\int_E f = 0$, *even if* $\mu(E) = +\infty$.

(*e*) If $\mu(E) = 0$, *then* $\int_E f = 0$, *even if* $f \equiv +\infty$ *everywhere on E.*

(*f*) $\int_E f = \int_R f\chi_E$. (*Cf. Exercise* 4–1.)

PROOF: Trivial. Note that (*f*) shows that any integral may be thought of as an integral over *R*.

Theorem 7: *If s and t are simple functions, then* $\int_E (s + t) = \int_E s + \int_E t$. (*From now on we avoid stating specifically the hypotheses that the functions and the sets with which we are dealing are measurable.*)

PROOF: Practically a carbon copy of proof of Theorem 4. (Note that the sum of simple functions is also simple.)

Theorem 8 (Monotone Convergence Theorem): *If* $0 \leqslant f_1 \leqslant f_2 \leqslant \cdots$, *and if* $f = \lim_{n\to\infty} f_n$, *then* $\int_A f = \lim_{n\to\infty} \int_A f_n$. (*Note that f is meaningful, and that it is measurable* [*by the corollary to Theorem* 4–6]; *also, by* (*a*) *of Theorem* 6, $\int_A f_1 \leqslant \int_A f_2 \leqslant \cdots$, *and so* $\int_A f$ *and* $\lim_{n\to\infty} \int_A f_n$ *both exist. The only problem is to show the equality.*)

PROOF: (*a*) By (*a*) of Theorem 6, $\int_A f_k \leqslant \int_A f$ for all indices *k*. Hence, $\int_A f \geqslant \lim_{k\to\infty} \int_A f_k$, and so it remains only to prove the reverse inequality, $\int_A f \leqslant \lim_{k\to\infty} \int_A f_k$. For convenience let us denote $\lim_{k\to\infty} \int_A f_k$ by the symbol α. (Incidentally, if $\alpha = +\infty$ there is nothing more to do, since $\int_A f \geqslant +\infty$ assures that $\int_A f = +\infty$. Thus, we may assume, if we wish, that $\alpha < +\infty$, but the following argument even covers the case $\alpha = +\infty$.)

(*b*) Choose any positive number *c* less than 1 and any measurable simple function *s* which is everywhere $\leqslant f$. (Remember that $s \geqslant 0$ everywhere.) Let $A_n = \{x \mid f_n(x) \geqslant cs(x)\}$. The A_n's are measurable and $A = A_1 \cup A_2 \cup \cdots$. (Why?) Using (*a*), (*b*), and (*c*) of Theorem 6, we obtain $\int_A f_n \geqslant \int_{A_n} f_n \geqslant \int_{A_n} (cs) = c\int_{A_n} s$, and so, letting $n \to \infty$, we obtain (noting that $\int_{A_n} s \leqslant \int_{A_{n+1}} s$)

$$\alpha \geqslant c \lim_{n\to\infty} \int_{A_n} s.$$

If $s =$ constant, $\int_{A_n} s =$ (constant value of $s)\mu(A_n) \to$ (constant value of $s)\mu(A)$ (by Theorem 3–13) $= \int_A s$; if s is *not* constant, a splitting argument similar to the one used in Theorem 4 shows that it is still true that $\int_{A_n} s \to \int_A s$. Thus, $\alpha \geqslant c \int_A s$. Since s is any simple function $\leqslant f$, by taking account of the definition of $\int_A f$ we obtain $\alpha \geqslant c \int_A f$. Now, letting $c \to 1$, we obtain $\alpha \geqslant \int_A f$, and so the proof is complete.

Theorem 9: *If f and g are non-negative, $\int_A (f + g) = \int_A f + \int_A g$. (Note that $f + g$ is measurable by Theorem 4–4; the $(+\infty) + (-\infty)$ difficulty cannot arise in this case.)*

PROOF: As in Theorem 1, construct a monotone sequence of simple functions $s_1, s_2, \ldots$ converging to f and a sequence $t_1, t_2, \ldots$ converging to g. Then the sequence $(s_1 + t_1), (s_2 + t_2), \ldots$ converges monotonely to $f + g$. By Theorems 7 and 8 we obtain

$$\begin{aligned} \int_A (f + g) &= \lim_{n\to\infty} \int_A (s_n + t_n) \\ &= \lim_{n\to\infty} \left\{ \int_A s_n + \int_A t_n \right\} \\ &= \lim_{n\to\infty} \int_A s_n + \lim_{n\to\infty} \int_A t_n = \int_A f + \int_A g. \end{aligned}$$

Theorem 10: *If the functions $f_1, f_2, \ldots$ are non-negative, then $\int_A \sum_{n=1}^{\infty} f_n = \sum_{n=1}^{\infty} \int_A f_n$. (By Theorem 4–4 and induction, $\sum_1^N f_n$ is measurable, and then by the corollary to Theorem 4–6 the infinite sum $\sum_1^{\infty} f_n$ is also measurable.)*

PROOF: Let $f(x) = \sum_{n=1}^{\infty} f_n(x)$ and let $g_n(x) = \sum_{k=1}^{n} f_k(x)$. Clearly, $g_n(x) \to f(x)$ everywhere and $0 \leqslant g_1 \leqslant g_2 \leqslant \cdots$. By Theorems 8 and 9 and finite induction,

$$\int_A f = \lim_{n\to\infty} \int_A g_n = \lim_{n\to\infty} \left\{ \int_A f_1 + \int_A f_2 + \cdots + \int_A f_n \right\},$$

and this is precisely the assertion of the theorem.

Theorem 11 (Fatou's Lemma): *If $f_1, f_2, \ldots$ are non-negative functions, then $\int_A \underline{\lim} f_k \leqslant \underline{\lim} \int_A f_k$.*

PROOF: Let $g_1 = \inf\{f_1, f_2, \ldots\}$, $g_2 = \inf\{f_2, f_3, \ldots\}$, and so forth. Clearly, $g_k \leqslant f_k$, and so $\int_A g_k \leqslant \int_A f_k$. Hence, $\underline{\lim} \int_A g_k \leqslant \underline{\lim} \int_A f_k$. However, since $g_1 \leqslant g_2 \leqslant g_3 \leqslant \cdots$, $\underline{\lim} \int_A g_k$ is actually $\lim_{k\to\infty} \int_A g_k$, and this, according to Theorem 8, is the same as $\int_A (\lim g_k)$. Hence, $\int_A (\lim g_k) \leqslant \underline{\lim} \int_A f_k$. From the definition of $\underline{\lim}$, we see that $\lim g_k = \underline{\lim} f_k$, and so the last inequality furnishes the desired result.

Exercise

1. Let $\{E_k\}$ be a sequence of measurable subsets of R and suppose that $\sum_{k=1}^{\infty} \mu(E_k)$ is finite. Prove that almost all points of R belong to only finitely many E_k's.

§6. INTEGRATION OF REAL-VALUED AND COMPLEX-VALUED FUNCTIONS

We now proceed to eliminate the restriction, which was imposed throughout the preceding section, that the functions involved may assume only non-negative values. First, we consider functions assuming values in R^*.

Definition 1: $f^+ = \max(f, 0)$, $f^- = \max(-f, 0)$. (*Note that f^- is a non-negative function, even though f^- is often called the negative part of f.) Obviously $f = f^+ - f^-$. Note that if f is measurable, so are f^+ and f^-.*

Definition 2: *Suppose that f is measurable and does not assume either of the values $\pm\infty$. (We shall later see that this restriction can be removed, but at this point it would be slightly bothersome to do so.) If $\int f^+$ and $\int f^-$ are both finite, we define $\int f$ to be $\int f^+ - \int f^-$, and we say that f is integrable, or summable (over the set on which we are integrating). Note that, if $f \geqslant 0$ everywhere, $f^- \equiv 0$, and the new definition of $\int f$ agrees with the original definition.*

Remark: Note that if f is integrable, so is $|f|$; in fact, $\int |f| = \int f^+ + \int f^-$. This is in marked contrast to *improper* Riemann integration; for example, in the Riemann theory we assign the value $\pi/2$ to $\int_0^{+\infty} \sin x/x$, since $\lim_{a\to+\infty} \int_0^a \sin x/x$ exists and equals $\pi/2$. Since $\int_0^\infty |\sin x/x| = +\infty$ (i.e., $\sin x/x$ is *not* summable), the statement $\int_0^\infty \sin x/x = \pi/2$ is *false* in the Lebesgue theory. (Cf. Exercise H-1.) (Since integrals of this type play an important role in Fourier analysis, we see that the Lebesgue theory is not quite a panacea!)

Theorem 1: *Let f and g be summable. Then (a) αf is summable for any real constant α, and $\int (\alpha f) = \alpha \int f$; (b) $f + g$ is summable, and $\int (f + g) = \int f + \int g$. (More generally, $\int (\alpha f + \beta g) = \alpha \int f + \beta \int g$.)*

PROOF: (*a*) Trivial.

(*b*) $f + g$ is measurable. Since $|f + g| \leqslant |f| + |g|$, it follows that $(f + g)^+$ and $(f + g)^-$ are non-negative and $\leqslant |f| + |g|$. Thus, by earlier results, $\int (f + g)^+ \leqslant \int (|f| + |g|) = \int |f| + \int |g| < +\infty$, and similarly for $(f + g)^-$. Thus, $f + g$ is *summable*. Let $h = f + g$ and form h^+ and h^-. (Note carefully that $h^+ = f^+ + g^+$ is *not* always true!!) Then

$$h^+ - h^- = f^+ - f^- + g^+ - g^-,$$

and so

$$h^+ + f^- + g^- = h^- + f^+ + g^+.$$

In this equation everything in sight is $\geqslant 0$, so by earlier results we may integrate term-by-term, obtaining

$$\int h^+ + \int f^- + \int g^- = \int h^- + \int f^+ + \int g^+,$$

and then we may transpose (since all quantities are finite!!):

$$\int h^+ - \int h^- = \left\{\int f^+ - \int f^-\right\} + \left\{\int g^+ - \int g^-\right\},$$

or

$$\int h = \int f + \int g,$$

or

$$\int (f + g) = \int f + \int g,$$

and the proof is complete.

Theorem 2: *If f is summable, $|\int f| \leqslant \int |f|$.*

PROOF: Since f^+ and f^- are non-negative, their *integrals* are $\geqslant 0$, and so

$$-\int |f| = -\int (f^+ + f^-) = -\int f^+ - \int f^- \leqslant \int f = \int f^+ - \int f^-$$
$$\leqslant \int f^+ + \int f^- = \int (f^+ + f^-) = \int |f|.$$

This is equivalent (disregarding the intermediate terms) to $|\int f| \leqslant \int |f|$, which is just the assertion of theorem.

Theorem 3 (Lebesgue Dominated Convergence Theorem): *Suppose that the functions $f_1, f_2, f_3, \ldots$ are measurable, that $|f_k(x)| \leqslant g(x)$*

everywhere, $g(x)$ being summable, and suppose that $f_n(x) \to f(x)$ everywhere. Then (a) f is summable; (b) $\int |f - f_n| \to 0$; (c) $\lim_{n\to\infty} \int f_n$ exists; and (d) the limit appearing in (c) coincides with $\int f$—that is, $\int \lim f_n = \lim \int f_n$.

PROOF: (*a*) By earlier results, f is measurable. Since each $|f_k(x)|$ is $\leqslant g(x)$, it follows that $|f(x)| \leqslant g(x)$, and so $f^+ \leqslant g$ and $f^- \leqslant g$ everywhere. Thus, $0 \leqslant \int f^\pm \leqslant \int g < +\infty$ (by (*a*) of Theorem 5–6), and so f is summable.

(*b*) The functions $2g - |f_n - f|$ are measurable and non-negative and converge to $2g$; in particular, $\underline{\lim}_{n\to\infty} \{2g - |f_n - f|\} \equiv 2g$, and by Fatou's lemma

$$\int 2g \leqslant \underline{\lim}_{n\to\infty} \int \{2g - |f_n - f|\}.$$

Now, $|f_n - f|$ is summable, since it is measurable, non-negative, and $\leqslant 2g$ (use (*a*) of Theorem 5–6), and so, by Theorem 1,

$$\int \{2g - |f_n - f|\} = \int 2g - \int |f_n - f|.$$

Since $2g$ is independent of the index n,

$$\underline{\lim}_{n\to\infty} \int \{2g - |f_n - f|\} = \underline{\lim}_{n\to\infty} \left\{\int 2g - \int |f_n - f|\right\}$$

$$= \int 2g - \overline{\lim}_{n\to\infty} \int |f_n - f|.$$

(Note carefully the change from $\underline{\lim}$ to $\overline{\lim}$, because of the minus sign!) Thus, $\int 2g \leqslant \int 2g - \overline{\lim}_{n\to\infty} \int |f_n - f|$; since $\int 2g$ is finite, we may cancel and transpose, obtaining

$$\overline{\lim}_{n\to\infty} \int |f_n - f| \leqslant 0.$$

But the quantities $\int |f_n - f|$ are $\geqslant 0$ by their very nature, and so $\underline{\lim}_{n\to\infty} \int |f_n - f| \geqslant 0 \geqslant \overline{\lim}_{n\to\infty} \int |f_n - f|$. Since $\underline{\lim} \leqslant \overline{\lim}$, we *must* have $\underline{\lim} = 0 = \overline{\lim} = \lim$. Thus, $\lim_{n\to\infty} \int |f_n - f| = 0$, which is (*b*).

(*c*) and (*d*): (*b*) guarantees that for every positive ϵ there exists $N(\epsilon)$ such that whenever $n > N(\epsilon)$ the inequality $\int |f - f_n| < \epsilon$ holds. By Theorem 2, $|\int (f - f_n)| < \epsilon$, and by Theorem 1 $|\int f - \int f_n| < \epsilon$. But this says that $\int f_n \to \int f$, which contains both (*c*) and (*d*).

It should be entirely evident now that summability, value of an integral, and so forth are not affected by altering a function on a set of measure zero; we may therefore change a function on a null-set, or disregard on a null-set, whenever we find it convenient. For example, in the Lebesgue dominated convergence theorem, we may replace the word "everywhere" in "$f_n(x) \to f(x)$ everywhere" and "$|f_n(x)| \leqslant g(x)$ everywhere" by "almost everywhere." In fact, we may speak of $\int_A f$ even if f is defined only almost everywhere on A; we simply integrate over A minus a suitable null-set, or else we may extend the definition of f to all of A in any way we please.

We conclude this section by indicating briefly how the entire theory of integration which we have developed is extended to complex-valued functions (defined on a subset of R). Let f be a complex-valued function defined on a set A, and let g and h be the real and imaginary parts of f, respectively. Then f is said to be measurable iff both g and h are measurable; similarly, f is said to be summable iff both g and h are summable, and when these conditions are satisfied the integral $\int_A f$ is defined as $\int_A g + i \int_A h$. The entire theory then extends, with obvious modifications, to the integration of complex-valued functions.

Exercises

1. Prove that a complex-valued function f is summable iff f is measurable and $|f|$ is summable.

2. (*a*) If f is real-valued and summable, prove that $|\int f| = \int |f|$ iff either $f \geqslant 0$ almost everywhere or $f \leqslant 0$ almost everywhere.
   (*b*) If f is complex-valued and summable, prove that $|\int f| = \int |f|$ iff there exists a constant c of unit modulus such that $f = c\,|f|$ almost everywhere.

3. If f is a summable function, either real-valued or complex-valued, defined on a set A, show that for every positive number ϵ there exists a positive number δ such that, whenever B is a subset of A whose measure is less than δ, the inequality $\int_B |f| < \epsilon$ (and hence the inequality $|\int_B f| < \epsilon$) holds. (Of course, this is trivial if $|f|$ is bounded.)

§7. INTEGRATION OVER PLANE SETS

In attempting to extend the theory of measure and integration from the real line, R, to the plane, $R \times R$, one encounters at the very beginning the difficulty that for open sets in the plane there is no canonical decomposition analogous to that which exists for open sets in R, as stated in §2 and explained in Appendix E. It is simply not true, for example, that every open subset of the plane can be expressed as the union of a finite or countably infinite collection of disjoint open squares (or rectangles). This becomes evident as soon as one considers a circular disc or the domain formed by removing a square from one corner of a larger square. Fortunately, however, a suitable decomposition theorem does exist.

Theorem 1: *Every non-empty open subset of the plane can be expressed as the union of a countably infinite collection of closed squares whose interiors are disjoint. Although this type of decomposition is not unique, the sum of the areas of the squares employed is the same for all decompositions.*

We shall merely sketch a proof of the first half of this theorem; for a complete proof of the entire theorem, and also of Theorem 2, the reader is referred to [Hartman-Mikusinski, Chapters 11 and 12]. A uniformly spaced mesh of horizontal and vertical lines is constructed, the spacing being sufficiently fine that at least one of the closed squares defined by the mesh is contained in the given open set O. To each such square the label "1" is attached. The mesh is then refined by constructing lines midway between each successive pair of lines of the original mesh. To those squares (if any exist) of the refined mesh which are contained in O but are not subsets of the squares already labeled, the label "2" is attached. As this procedure is repeated indefinitely, a countably infinite collection of closed squares is obtained whose interiors are disjoint and whose union is precisely the given open set O. In Figure 1 the first three stages of the decomposition are illustrated.

The two parts of the preceding theorem, taken together, justify the assignment to every plane open set O of a number (possibly $+\infty$) $\mu(O)$, defined as the sum of the areas of the squares used in any decomposition

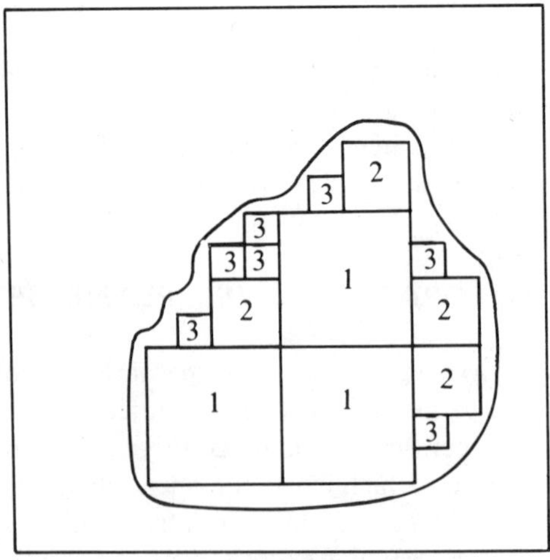

Figure 1.

of O of the type described in the statement of the theorem. It is not difficult to show that if O is a square (or rectangle), then $\mu(O)$ coincides with the area of O. (Cf. Exercise 1.) It is now possible to repeat the entire chain of arguments employed in the preceding sections of this chapter to obtain a satisfactory theory of measure and integration in the plane. The integral of a (summable) function f over a (measurable) set A will be denoted $\iint_A f$, or $\iint_A f(x, y)\, dx\, dy$.

The following question and its answer are of fundamental significance in many problems of analysis: If f is summable over the rectangle A, defined by the inequalities* $a < x < b$ and $c < y < d$, is the value of the integral $\iint_A f$ equal to that of the iterated integral $\int_c^d \{\int_a^b f(x, y)\, dx\}\, dy$ and of the iterated integral $\int_a^b \{\int_c^d f(x, y)\, dy\}\, dx$? In fact, it is far from obvious that the existence of the plane integral guarantees the existence of either iterated integral. To take a rather trivial example, let C be a non-measurable subset of the interval $(-1, 1)$, let A be the open square $(-1, 1) \times (-1, 1)$, and let f be defined in A as the characteristic function of the point-set $\tilde{C}$ defined by the pair of conditions $x \in C$, $y = 0$. It is extremely important to note that, while C is a non-measurable subset of R, $\tilde{C}$ is a measurable subset of $R \times R$—in fact, $\mu(\tilde{C}) = 0$. Thus $\iint_A f = 0$; also, $\int_{-1}^{1} f(x, y)\, dy = 0$ for all x in $(-1, 1)$, and hence

$$\int_{-1}^{1}\left\{\int_{-1}^{1} f(x, y)\, dy\right\} dx = 0.$$

On the other hand, $\int_{-1}^{1} f(x, y)\, dx = 0$ if $y \neq 0$, but $\int_{-1}^{1} f(x, 0)\, dx$ does not exist (since C is not measurable). However, since $\int_{-1}^{1} f(x, y)\, dx$ is defined for almost all values of y and is summable, the iterated integral $\int_{-1}^{1} \{\int_{-1}^{1} f(x, y)\, dx\}\, dy$ exists and equals zero. This example, simple though it is, serves to indicate the difficulties which must be overcome in establishing the following celebrated theorem.

Theorem 2 (Fubini): *Let f be summable over the rectangle $A = (a, b) \times (c, d)$. Then $\int_a^b f(x, y)\, dx$ exists for almost all values of y in (c, d), and the function defined by this integral is summable, so that the iterated integral $\int_c^d \{\int_a^b f(x, y)\, dx\}\, dy$ exists; similarly, the iterated integral $\int_a^b \{\int_c^d f(x, y)\, dy\}\, dx$ exists. The plane integral $\iint_A f$ and the two iterated integrals all have the same value.*

Exercise 2 shows that the existence and equality of the two iterated integrals does not guarantee the existence of the plane integral, while Exercise 3 shows that it is possible that only one of the two iterated integrals may exist. Finally, Exercise 4 shows that both iterated integrals may exist and yet have different values.

---

* Nothing is altered by replacing one or more of the strict inequalities by $\leqslant$. Also, a and c may equal $-\infty$ while b and d may equal $+\infty$.

Exercises

1. Let A be the open rectangle $(a, b) \times (c, d)$. Show that $\mu(A) = (b - a) \cdot (d - c)$, as is to be expected.

2. Let A be the open square $(-1, 1) \times (-1, 1)$ and let
$$f(x, y) = \frac{xy}{(1 - |x|)^2 + (1 - |y|)^2}.$$
Show that the two iterated integrals exist and equal zero, but that $\iint_A f$ does not exist.

3. Let A be the open square $(-1, 1) \times (-1, 1)$ and let $f(x, y) = x/(1 - y^2)$. Show that $\int_{-1}^{1} \{\int_{-1}^{1} f(x, y)\, dx\}\, dy$ exists and equals zero, while $\int_{-1}^{1} \{\int_{-1}^{1} f(x, y)\, dy\}\, dx$ does not exist.

4. Let A be the open square $(0, 1) \times (0, 1)$ and let $f(x, y) = y^{-3/2}$, 0, or $-y^{-1/2}(1 - y)^{-1}$ according as $y > x$, $y = x$, or $y < x$. Show that both iterated integrals exist but that their values are different.

§8. CONCLUDING REMARKS

We have now completed the development of the essential parts of the theory of Lebesgue measure and integration, at least to the extent that we shall need it. In this section we present a few remarks that may prove of interest and value to the reader.

(A) Suppose that f is a continuous function defined on a bounded closed interval $[a, b]$. Then it is very easy to show that $\int_a^b f(x)\, dx$ has the same value whether interpreted as a Riemann integral or as a Lebesgue integral. We shall use this fact occasionally without specific mention. By a slightly more delicate argument one can show that the aforementioned equality holds if f is any function which is integrable in the sense of Riemann. (There is no contradiction, of course, between this statement and the remark following Definition 6–2, since the integral mentioned there is extended over an infinite interval.) It is also of interest to note the following theorem, which demonstrates a remarkable connection between Riemann integration and Lebesgue measure: A function f defined on a bounded closed interval is integrable in the sense of Riemann iff $|f|$ is bounded and the points of discontinuity of f form a null-set.

(B) The entire theory can be extended in the following way. Let a non-empty set X be given, and let a *sigma-algebra* (also called *Borel field*) of subsets of X be given. This means that we have a non-empty class $\mathscr{B}$ (for "Borel") of subsets of X which satisfy the following conditions: (a) The complement of any member of $\mathscr{B}$ is also a member of $\mathscr{B}$; (b) The union of finitely many or countably many members of $\mathscr{B}$ is also a member of $\mathscr{B}$; (*c*) $\emptyset$ and X are members of $\mathscr{B}$; (*d*) The intersection of finitely many or countably many members of $\mathscr{B}$ is also a member of $\mathscr{B}$. (In fact, (*c*) and (*d*) are consequences of (*a*) and (*b*), but for simplicity we have included them.) Furthermore, suppose that to each set A belonging to $\mathscr{B}$ we associate a non-negative number ($+\infty$ allowed), $\mu(A)$, satisfying the following conditions: (*a*) $\mu(\emptyset) = 0$; (*b*) $\mu(\bigcup A_k) = \sum \mu(A_k)$ for any *finite* or *countable disjoint* union of members of $\mathscr{B}$; (*c*) if $A_1 \subseteq A_2$, if $A_2 \in \mathscr{B}$, and if $\mu(A_2) = 0$, then $A_1 \in \mathscr{B}$ and $\mu(A_1) = 0$. (In fact, $\mu(A_1) = 0$ is a consequence of the other assumptions.)

Such a function μ is called a measure on the class $\mathscr{B}$, and the entire theory which we have built up remains valid in this more general context. In particular, probability theory depends in an essential way on the theory of measures defined on Borel fields.

(Strictly speaking, the condition that every subset of a set of measure zero should belong to $\mathscr{B}$ is not imposed in the definition of a measure; when this condition is satisfied the measure is said to be complete. However, every incomplete measure defined on a Borel field can be extended to a larger Borel field so that the new measure is complete.)

CHAPTER 3

THE L^p- AND l^p-SPACES

All functions under discussion in the first five sections of this chapter are complex-valued and defined on a fixed measurable subset A of R. We assume that $\mu(A)$ is positive ($+\infty$ allowed), for if $\mu(A) = 0$ the entire content would become vacuous. In the first four sections the number p stands for a fixed *finite* number $\geqslant 1$, and in §5 it will be indicated briefly how the material developed in the first four sections can be extended to the case $p = +\infty$.

§1. BASIC CONCEPTS

Definition 1: *Suppose that f is measurable on A and that $|f|^p$ is summable over A. (Note carefully that the measurability of f guarantees the measurability, but not the summability, of $|f|^p$. Cf. Exercises* 1 *and* 2.) *Then f is called p-th power summable over A. (When $p = 1$ this is the same as summable; when $p = 2$ we use the expressions square integrable or quadratically integrable.)*

Definition 2: *The class of all p-th power summable functions is denoted $L^p(A)$; since A is fixed, we usually employ the simpler notation L^p. If $f \in L^p$, the non-negative finite number $(\int_A |f|^p)^{1/p}$ is denoted $\|f\|_p$. (From now on we write $\int$ instead of $\int_A$.)*

Theorem 1: (*a*) *If α is any complex number and if $f \in L^p$, then $\alpha f \in L^p$ and $\|\alpha f\|_p = |\alpha| \cdot \|f\|_p$.*

(*b*) $\|f\|_p = 0$ *iff $f(x) = 0$ almost everywhere.*

PROOF: Trivial.

The following theorem, despite its simplicity, plays a vital role in the theory to be developed later in this chapter, and we present two quite distinct proofs.

Theorem 2: *Let x and y be non-negative real numbers and let $0 < r < 1$. Then $x^r y^{1-r} \leqslant rx + (1 - r)y$, and equality holds iff $x = y$.*

FIRST PROOF: Since the assertions of this theorem are obvious if either x or y vanishes, we confine attention to the case $x > 0, y > 0$. Let a and b be positive numbers. Then $0 \leqslant (\sqrt{a} - \sqrt{b})^2 = a + b - 2\sqrt{ab}$, and so $a^{1/2}b^{1/2} \leqslant \frac{1}{2}(a + b)$. (This is, of course, the familiar inequality between the geometric and arithmetic means.) Replacing a by $c_1^{1/2}c_2^{1/2}$ and b by $c_3^{1/2}c_4^{1/2}$, where the c's are arbitrary positive numbers, we obtain $c_1^{1/4}c_2^{1/4}c_3^{1/4}c_4^{1/4} \leqslant \frac{1}{2}(c_1^{1/2}c_2^{1/2} + c_3^{1/2}c_4^{1/2}) \leqslant \frac{1}{2}\{\frac{1}{2}[c_1 + c_2] + \frac{1}{2}[c_3 + c_4]\} = \frac{1}{4}(c_1 + c_2 + c_3 + c_4)$. Repeating this procedure any finite number of times we obtain, for any positive integer n and positive numbers $\alpha_1, \alpha_2, \ldots, \alpha_{2^n}$,

$$(\alpha_1\alpha_2 \cdots \alpha_{2^n})^{1/2^n} \leqslant \frac{1}{2^n}(\alpha_1 + \alpha_2 + \cdots + \alpha_{2^n}).$$

If we set k of the α's equal to x and the remaining α's equal to y (where $0 < k < 2^n$), we obtain

$$x^{k/2^n}y^{1-k/2^n} \leqslant \frac{k}{2^n}x + \left(1 - \frac{k}{2^n}\right)y.$$

Thus, the desired inequality has been established in the particular case that r is a *binary rational*—a rational number which can be written as a fraction whose denominator is a power of 2—in the open interval (0, 1). Since the binary rationals are dense in R, we now conclude by a trivial continuity argument that the desired inequality holds for *any* value of r in (0, 1). Finally, by working backwards through the entire argument, we see that equality holds iff $x = y$, as asserted.

SECOND PROOF: Let the function f be defined as follows: $f(t) = t^r - 1 + r - rt$ for $t \geqslant 0$. Then f is continuous, while for positive values of t it possesses derivatives of all orders. In particular, $f'(t) = rt^{r-1} - r$ and $f''(t) = r(r - 1)t^{r-2} < 0$. The latter equation shows that the function f is concave, and from the first equation it is then seen that $t = 1$, and no other value of t, furnishes the maximum value of f. Thus, $f(t) < f(1)$, or $t^r < (1 - r) + rt$, for t positive and different from unity. Replacing t by x/y, where x and y are distinct positive numbers, we obtain $x^r y^{-r} < (1 - r) + rx/y$, or $x^r y^{1-r} < rx + (1 - r)y$, while equality must hold when $x = y$.

Definition 3: *If $1 < p < +\infty$, the conjugate number of p is the number $p/(p-1)$; we always denote the conjugate of p by q. (Since $1/p + 1/q = 1$, we see that p is the conjugate number of q. Note that 2 is its own conjugate, and that as $p \to \begin{Bmatrix} 1 \\ +\infty \end{Bmatrix}$ its conjugate $\to \begin{Bmatrix} +\infty \\ 1 \end{Bmatrix}$. We therefore sometimes find it convenient to consider 1 and $+\infty$ as the conjugates of each other.)*

Exercises

1. Prove that the measurability of f guarantees the measurability of $|f|^p$.
2. (*a*) Give an example of a function belonging to $L^1([0, 1])$ but not belonging to $L^2([0, 1])$.
   (*b*) Determine a set A and a function belonging to $L^2(A)$ but not belonging to $L^1(A)$. (As will become clear in §2, $\mu(A)$ must be $+\infty$ in this case.)

§2. THE HÖLDER AND MINKOWSKI INEQUALITIES

Theorem 1: *If $f \in L^p$ and $g \in L^q$, then $fg \in L^1$.*

PROOF: Let $x = |f|^p$, $y = |g|^q$, and $r = 1/p$ (so that $1 - r = 1/q$), and substitute these values into Theorem 1–2. We obtain $|f| \cdot |g| \leqslant 1/p\,|f|^p + 1/q\,|g|^q$. Since $|fg|$ $(= |f| \cdot |g|)$ is the product of two measurable functions, it is also measurable, and since $1/p\,|f|^p + 1/q\,|g|^q$ is *summable*, so is $|fg|$. We then know that fg is summable—i.e., $fg \in L^1$.

Theorem 2 (Hölder Inequality): *If $f \in L^p$ and $g \in L^q$, then $\|fg\|_1 \leqslant \|f\|_p \|g\|_q$.*

PROOF: (*a*) By (*b*) of Theorem 1–1, the assertion is trivially true if either $\|f\|_p = 0$ or $\|g\|_q = 0$.

(*b*) Suppose $\|f\|_p = 1$ and $\|g\|_q = 1$. Integrating the inequality developed in Theorem 1, we obtain

$$\|fg\|_1 = \int |fg| \leqslant \frac{1}{p}\int |f|^p + \frac{1}{q}\int |g|^q = \frac{1}{p} + \frac{1}{q} = 1 = \|f\|_p \|g\|_q.$$

(*c*) On account of (*a*), we may assume that $\|f\|_p > 0$ and $\|g\|_q > 0$. Let $f_1 = (1/\|f\|_p)f$ and $g_1 = (1/\|g\|_q)g$. By (*a*) of Theorem 1–1, $\|f_1\|_p = \|g_1\|_q = 1$, and now, by (*b*) of the present proof,

$$\|f_1 g_1\|_1 \leqslant 1, \quad \text{or} \quad \left\| \frac{1}{\|f\|_p \|g\|_q} fg \right\|_1 \leqslant 1.$$

Again referring to (*a*) of Theorem 1–1, we obtain, as desired,

$$\|fg\|_1 \leqslant \|f\|_p \, \|g\|_q.$$

Theorem 3: *If f and $g \in L^p$, then $f + g \in L^p$.*

PROOF:

$$|f+g|^p \leqslant (|f| + |g|)^p \leqslant (2 \max \{|f|, |g|\})^p$$
$$= 2^p \max \{|f|^p, |g|^p\} \leqslant 2^p (|f|^p + |g|^p).$$

(The last inequality follows from the obvious fact that the larger of two non-negative quantities is at most equal to their sum.) Hence $|f+g|^p$ is summable, and so $f + g \in L^p$.

Corollary: *If $f_1, f_2, \ldots, f_n \in L^p$, then any linear combination of these functions, $\alpha_1 f_1 + \alpha_2 f_2 \cdots + \alpha_n f_n$, also belongs to L^p.*

PROOF: Use Theorem 3, (*a*) of Theorem 1–1, and finite induction.

Theorem 4 (Minkowski Inequality): *If f and $g \in L^p$, then $\|f+g\|_p \leqslant \|f\|_p + \|g\|_p$. (By Theorem 3 we know that the left side of this inequality is meaningful.)*

PROOF: (*a*) If $p = 1$, we argue as follows: $|f+g| \leqslant |f| + |g|$, hence $\int |f+g| \leqslant \int |f| + \int |g|$, or $\|f+g\|_1 \leqslant \|f\|_1 + \|g\|_1$.

(*b*) If $p > 1$, we argue as follows: $|f+g|^p = |f+g| \cdot |f+g|^{p-1} \leqslant (|f| + |g|) \cdot |f+g|^{p-1} = |f| \cdot |f+g|^{p-1} + |g| \cdot |f+g|^{p-1}$. Thus,

$$\int |f+g|^p \leqslant \int |f| \cdot |f+g|^{p-1} + \int |g| \cdot |f+g|^{p-1}.$$

In Theorem 2 replace f by $|f|$ and g by $|f+g|^{p-1}$; also note that $\|f\|_p = \| \, |f| \, \|_p$ and $\|f+g\|_q = \| \, |f+g| \, \|_q$. We obtain (using the fact that $(p-1)q = p$) $\int |f| \cdot |f+g|^{p-1} \leqslant \|f\|_p \cdot \| \, |f+g|^{p-1} \|_q = \|f\|_p \cdot (\int |f+g|^{(p-1)q})^{1/q} = \|f\|_p \cdot (\int |f+g|^p)^{1/q} = \|f\|_p \cdot \|f+g\|_p^{p/q}$. Similarly, $\int |g| \cdot |f+g|^{p-1} \leqslant \|g\|_p \cdot \|f+g\|_p^{p/q}$. Hence, $\int |f+g|^p \leqslant (\|f\|_p + \|g\|_p) \cdot (\|f+g\|_p)^{p/q}$, or $\|f+g\|_p^p \leqslant (\|f\|_p + \|g\|_p) \cdot (\|f+g\|_p)^{p/q}$. Dividing by $(\|f+g\|_p)^{p/q}$, we obtain $(\|f+g\|_p)^{p-p/q} \leqslant \|f\|_p + \|g\|_p$. But $p - p/q = p(1 - 1/q) = p \cdot 1/p = 1$. Hence, $\|f+g\|_p \leqslant \|f\|_p + \|g\|_p$. (If $\|f+g\|_p = 0$, the division is not legitimate, but in this case no proof is needed.)

Exercises

1. Prove the Hölder and Minkowski inequalities in the particular case $p = q = 2$ by exploiting the fact that $\int |f + \lambda g|^2 \geqslant 0$ for all values of the real parameter λ.

2. (*a*) Assuming that f and g never vanish on A, prove that $\|fg\|_1 = \|f\|_p \|g\|_q$ iff the fractions $|f|^p/|g|^q$ and $fg/|fg|$ are both constant almost everywhere on A.

(*b*) Assuming that $\|f\|_p$ and $\|g\|_p$ are both positive, show that $\|f+g\|_p = \|f\|_p + \|g\|_p$ iff there exists a positive constant α such that $f(x) = \alpha g(x)$ almost everywhere.

3. In the inequality $|fg| \leqslant (1/p)|f|^p + (1/q)|g|^q$ which is employed in Theorem 1, replace f by cf and g by g/c, where c is an unspecified positive constant. The preceding inequality becomes

$$|fg| \leqslant \frac{c^p}{p}|f|^p + \frac{c^{-q}}{q}|g|^q,$$

and by integration we obtain

$$\|fg\|_1 \leqslant \frac{c^p}{p}\|f\|_p^p + \frac{c^{-q}}{q}\|g\|_q^q.$$

Show how to obtain the general form of the Hölder inequality directly from this result.

§3. DEFINITION OF A METRIC IN L^p

For any two functions f and g belonging to L^p, let us define $\rho(f, g)$ as $\|f - g\|_p$. (According to Theorem 2–3, $\|f - g\|_p$ is finite; we merely replace g by $-g$.) Clearly, $\rho(f, g) = \rho(g, f)$, and from Theorem 2–4 we obtain

$$\rho(f, h) = \|f - h\|_p = \|(f - g) + (g - h)\|_p$$
$$\leqslant \|f - g\|_p + \|g - h\|_p = \rho(f, g) + \rho(g, h).$$

Referring back to Chapter 1, we see that the second and third axioms of a metric space are satisfied. However, a slight complication arises in connection with the first axiom, for the vanishing of $\rho(f, g)$ does not guarantee that f and g are the same function—according to Theorem 1–1, f and g may differ on a null-set (but not on any larger set). This observation suggests how to salvage the situation. Instead of considering L^p to consist of functions (defined on A) as we have done until now, we form equivalence classes of these functions, two functions being considered equivalent iff they coincide almost everywhere on A. It is easily shown that this is indeed an equivalence relation and that the resulting collection of equivalence classes becomes a metric space (which we continue to denote L^p) if the "distance" between any two equivalence classes, E_1 and

E_2, is defined as $\|f_1 - f_2\|_p$, where f_1 and f_2 are *any* members of E_1 and E_2, respectively. (Cf. Exercise 1.)

We continue to use such expressions as "consider the function f in L^p"; but this must now be understood as a condensed version of the statement "consider the class of all p-th power summable functions which differ from f only on a null-set."

An alternative device, which avoids the need of forming equivalence classes, is to consider L^p as a *pseudo-metric* space, in which, while continuing to require that $\rho(f, g) \geqslant 0$ for all elements f and g, we permit $\rho(f, g) = 0$ to hold even though $f \neq g$. Which of these two schemes is chosen seems to be entirely a matter of taste; mathematically, they are almost certainly of equal value.

We conclude this section with the remark that the distance function which has been defined is translation-invariant. By this we mean that for any three functions f, g, h belonging to L^p, the distance between f and g is the same as the distance between $f + h$ and $g + h$. In the following chapter we shall deal with a class of metric spaces in which the distance function possesses this important property.

Exercise

1. Justify in detail the assertion that the "new" L^p is indeed a metric space.

§4. COMPLETENESS OF L^p

We recall that a metric space M is said to be complete if every Cauchy sequence $x_1, x_2, \ldots$ in M is convergent—that is, there exists a member x of M such that $\lim_{n\to\infty} \rho(x_n, x) = 0$. The following theorem, which asserts that L^p is complete (in fact, it asserts even more than this), is justly regarded as one of the great theorems of analysis, and in particular one of the highlights of the Lebesgue theory.

Theorem I (Riesz-Fischer): *Let $f_1, f_2, \ldots$ be a Cauchy sequence in L^p. Then there exists a function f in L^p such that $\|f - f_n\|_p \to 0$. Furthermore, although the sequence may fail to converge (pointwise) anywhere in A, it is always possible to select a subsequence of the given sequence which converges almost everywhere in A to f.*

PROOF: (*a*) Taking account of Exercise 4–4 of Chapter 1, we see that it suffices to prove that we can select a subsequence of the given sequence, say $\tilde{f}_1, \tilde{f}_2, \ldots$, and a function f in L^p such that $\|f - \tilde{f}_n\|_p \to 0$ and $\tilde{f}_n(x) \to f(x)$ for almost all points x in A.

(*b*) Since the sequence $f_1, f_2, \ldots$ is Cauchy, we can choose an index N_1 so large that $\|f_n - f_m\| < 1/2$ whenever n and $m > N_1$. Then we can choose N_2 larger than N_1 such that $\|f_n - f_m\| < 1/2^2$ whenever n and $m > N_2$, and we continue this procedure. ($\| \ \|$ means $\| \ \|_p$; the numbers $1/2, 1/2^2, 1/2^3, \ldots$ can be replaced by any other sequence of positive numbers whose sum is finite.) Then we can choose a subsequence $\tilde{f}_1, \tilde{f}_2, \tilde{f}_3, \ldots$ in such a manner that $\|\tilde{f}_1 - \tilde{f}_2\| < 1/2$, $\|\tilde{f}_2 - \tilde{f}_3\| < 1/2^2$, etc. (For example, let $\tilde{f}_1 = f_{N_1+1}$, $\tilde{f}_2 = f_{N_2+1}$, etc.) Let $g_1 = |\tilde{f}_1 - \tilde{f}_2|$, $g_2 = |\tilde{f}_2 - \tilde{f}_3|$, etc. Since $\|g_k\| = \|\tilde{f}_k - \tilde{f}_{k+1}\| < 1/2^k$, we obtain, by the triangle inequality,

$$\|g_1 + g_2 + \cdots + g_k\| < \frac{1}{2} + \frac{1}{2^2} + \cdots + \frac{1}{2^k} < 1.$$

Let $s_n = \sum_{k=1}^n g_k$. The s_n's are non-negative, they form a monotone non-decreasing sequence, and their norms are bounded above by unity. By the monotone convergence theorem, there exists a function τ belonging to L^p such that $\int \tau^p \leqslant 1$ and $s_m \to \tau(x)$ everywhere; of course, $\tau(x)$ is finite almost everywhere. Therefore, the series $\sum_{k=1}^\infty |\tilde{f}_k - \tilde{f}_{k+1}|$ converges to a finite sum almost everywhere; *a fortiori*, the series $\sum_{k=1}^\infty (\tilde{f}_k - \tilde{f}_{k+1})$ converges almost everywhere to a function h such that $\|h\| \leqslant \|\tau\| \leqslant 1$. Since the partial sum $\sum_{k=1}^n (\tilde{f}_k - \tilde{f}_{k+1})$ telescopes to $\tilde{f}_1 - \tilde{f}_{n+1}$, we conclude that $\lim_{n\to\infty} \tilde{f}_{n+1}$ exists finitely almost everywhere and equals $\tilde{f}_1 - h$, which, being the sum of two members of L^p, is also a member of L^p. We have therefore shown that $\tilde{f}_n$ converges pointwise almost everywhere to the function $f = \tilde{f}_1 - h$, which belongs to L^p.

(*c*) Now we shall show that $\|f - \tilde{f}_n\| \to 0$. From (*b*) it is clear that $|\tilde{f}_n - \tilde{f}_1| \leqslant \tau$ and $|f - \tilde{f}_1| \leqslant \tau$. Hence,

$$|f - \tilde{f}_n| = |(f - \tilde{f}_1) + (\tilde{f}_1 - \tilde{f}_n)| \leqslant |f - \tilde{f}_1| + |\tilde{f}_n - \tilde{f}_1| \leqslant 2\tau,$$

and so $|f - \tilde{f}_n|^p \leqslant 2^p\tau^p$. Since τ^p is summable and since $|f - \tilde{f}_n|^p \to 0$ almost everywhere, we may apply the Lebesgue dominated convergence theorem to the sequence of functions $|f - \tilde{f}_1|^p, |f - \tilde{f}_2|^p, \ldots$. Thus, $\lim_{n\to\infty} \int |f - \tilde{f}_n|^p = \int \lim_{n\to\infty} |f - \tilde{f}_n|^p = \int 0 = 0$, and, therefore, $\|f - \tilde{f}_n\| \to 0$.

We now extend the preceding results a little further. Suppose that a *different* subsequence, say $\tilde{\tilde{f}}_1, \tilde{\tilde{f}}_2, \tilde{\tilde{f}}_3, \ldots$ converges almost everywhere, say to a function g. By repeating the preceding arguments we see that $\|g - f_n\| \to 0$; then, since $\|f - g\| \leqslant \|f - f_n\| + \|g - f_n\|$, we see that $\|f - g\| = 0$, and so $f = g$ almost everywhere. Thus, we have obtained the following result.

Corollary: *If a Cauchy sequence in L^p converges in the metric of L^p to f and pointwise almost everywhere to g, then f and g are equivalent*—i.e., $f(x) = g(x)$ *almost everywhere.*

It may be advisable to emphasize that the considerations of this section are valid for $p \geqslant 1$, not just for $p > 1$.

Exercise

1. (*a*) Show that for any positive integers m and n, the following equalities are correct:

$$(i)\quad \int_0^{2\pi} \sin^2 nx = \int_0^{2\pi} \cos^2 nx = \pi;$$

$$(ii)\quad \int_0^{2\pi} \sin nx \sin mx = \int_0^{2\pi} \cos nx \cos mx = 0 \text{ if } m \neq n;$$

$$(iii)\quad \int_0^{2\pi} \sin mx \cos nx = 0, \text{ even if } m = n.$$

(*b*) Let $f \in L^2([0, 2\pi])$. Show that the integrals $\int_0^{2\pi} f(x) \cos nx$ and $\int_0^{2\pi} f(x) \sin nx$ exist for $n = 1, 2, 3, \ldots$.

(*c*) Let the sequences $a_1, a_2, a_3, \ldots$ and $b_1, b_2, b_3, \ldots$ of complex numbers be given and suppose that the series $\sum_{k=1}^\infty |a_k|^2$ and $\sum_{k=1}^\infty |b_k|^2$ both converge (to *finite* sums). Show that there exists a function f in $L^2([0, 2\pi])$ such that all the equalities $a_k = 1/\pi \int_0^{2\pi} f(x) \cos kx$, $b_k = 1/\pi \int_0^{2\pi} f(x) \sin kx$ hold. Hint: Form the sequence of functions $f_1, f_2, f_3, \ldots$, where

$$f_n(x) = \sum_{k=1}^n (a_k \cos kx + b_k \sin kx).$$

§5. THE SPACE L^∞

How should we interpret L^p when $p = +\infty$? While we could simply give a definition, we shall try to *motivate* the answer. For simplicity, suppose $\mu(A)$ is finite (but, of course, positive). If $f \in L^p$ and if $|f| \leqslant C$ everywhere on A, then $\int |f|^p \leqslant \int C^p = C^p\mu(A)$, and so $\|f\|_p \leqslant C(\mu(A))^{1/p}$. As $p \to +\infty$, $(\mu(A))^{1/p} \to 1$, and so $\overline{\lim}_{p\to+\infty} \|f\|_p \leqslant C$. (Note that if f is measurable and bounded and $\mu(A)$ is *finite*, then $f \in L^p$ for all values of p.) On the other hand, suppose $|f| \geqslant D$ on some set B of *positive* measure. Then $\int_A |f|^p \geqslant \int_B |f|^p \geqslant \int_B D^p = D^p\mu(B)$, and so $\|f\|_p \geqslant D(\mu(B))^{1/p}$. Letting $p \to +\infty$, we obtain $\underline{\lim}_{p\to\infty} \|f\|_p \geqslant D$, since $(\mu(B))^{1/p} \to 1$. Putting together these two results, we see that $\lim_{p\to+\infty} \|f\|_p$ exists iff f is equivalent to a bounded function; of course, this includes the case that f itself is bounded. When this condition is satisfied f is said to be *essentially bounded.* A little thought will show that when the aforementioned limit of $\|f\|_p$ exists it is equal to the quantity $\inf_{g\sim f} \sup |g|$. (Recall that $g \sim f$ means that $g = f$ almost everywhere.) This quantity is called the *essential supremum* of $|f|$, denoted ess sup $|f|$; by its very definition it is the same for two equivalent functions. (If f is not essentially bounded, then it is not difficult to see that $\|f\|_p \to +\infty$. The reader is asked to justify these statements in Exercise 1.)

Therefore, we *define* L^∞ to be the set of all essentially bounded functions (more precisely, the set of equivalence classes of all measurable essentially bounded functions), and

$$\|f\|_\infty = \text{ess sup } |f|.$$

The reader should have no difficulty in seeing that all the results obtained in the earlier sections of this chapter continue to hold for L^∞, provided that the corresponding value of q is taken as unity; in fact, most of the proofs, particularly that of the Riesz-Fischer theorem, are easier for $p = +\infty$ than for $p < +\infty$.

Incidentally, note carefully that, although the *motivation* for our definition of L^∞ required that $\mu(A)$ should be finite, the *definition* is still applicable when $\mu(A) = +\infty$.

Exercise

1. Prove the assertions made concerning the behavior of $\|f\|_p$ as $p \to +\infty$ (under the assumption that $\mu(A)$ is finite).

§6. THE SPACES l^p_n AND l^p

Let n be an arbitrary but fixed positive integer and let $1 \leqslant p < +\infty$. For each pair of ordered n-tuples of complex numbers,

$$a = (\alpha_1, \alpha_2, \ldots, \alpha_n)$$

and $b = (\beta_1, \beta_2, \ldots, \beta_n)$, we define $\rho_p(a, b)$ as $\{\sum_{k=1}^n |\alpha_k - \beta_k|^p\}^{1/p}$. Clearly $\rho_p(a, b) = \rho_p(b, a)$, $\rho_p(a, b) > 0$ if $a \neq b$, and $\rho_p(a, b) = 0$ if $a = b$. In order to prove that ρ_p defines a metric on the collection of all ordered n-tuples it therefore suffices to show that the triangle inequality holds. To accomplish this, we let A be the subset $[0, n)$ of R and associate with the n-tuples a and b the functions $\tilde{a}$ and $\tilde{b}$, respectively, where $\tilde{a}(x) = \alpha_k$ when $x \in [k-1, k)$, $k = 1, 2, \ldots, n$, and similarly for $\tilde{b}$. It is obvious that $\tilde{a}$, $\tilde{b}$, and $\tilde{a} - \tilde{b}$ belong to $L^p(A)$, and that $\|\tilde{a} - \tilde{b}\|_p = \rho_p(a, b)$. If c denotes a third n-tuple, $(\gamma_1, \gamma_2, \ldots, \gamma_n)$, and if $\tilde{c}$ denotes the corresponding member of $L^p(A)$, we obtain from Theorem 2–4 the following: $\rho_p(a, b) = \|(\tilde{a} - \tilde{c}) + (\tilde{c} - \tilde{b})\|_p \leqslant \|\tilde{a} - \tilde{c}\|_p + \|\tilde{c} - \tilde{b}\|_p = \rho_p(a, c) + \rho_p(c, b)$, or, more explicitly,

$$\left\{\sum_{k=1}^n |\alpha_k - \beta_k|^p\right\}^{1/p} \leqslant \left\{\sum_{k=1}^n |\alpha_k - \gamma_k|^p\right\}^{1/p} + \left\{\sum_{k=1}^n |\gamma_k - \beta_k|^p\right\}^{1/p}. \quad (6\text{–}1)$$

Thus, we have shown that the collection of all ordered n-tuples of complex numbers is a metric space if the distance-function ρ_p previously defined is employed. (In particular, for $n = 3$ this result serves to complete the discussion of Example (m) of §1–2.) The metric space thus defined is denoted l_n^p. As in §5, we may let p approach $+\infty$, and we then obtain the sup metric: $\rho_\infty(a, b) = \sup_{1\leqslant k\leqslant n} \{|\alpha_k - \beta_k|\}$. The corresponding metric space is denoted, of course, by the symbol l_n^∞.

If $1 < p < +\infty$ and q is the conjugate number of p, the application of Theorem 2–2 to the functions $\tilde{a}$ and $\tilde{b}$ defined previously leads immediately to the inequality

$$\sum_{k=1}^{n} |\alpha_k\beta_k| \leqslant \left\{\sum_{k=1}^{n} |\alpha_k|^p\right\}^{1/p} \cdot \left\{\sum_{k=1}^{n} |\beta_k|^q\right\}^{1/q}. \tag{6–2}$$

Again we may let p approach $+\infty$; this leads to the trivial inequality

$$\sum_{k=1}^{n} |\alpha_k\beta_k| \leqslant \left\{\sup_{1\leqslant k\leqslant n} |\alpha_k|\right\} \cdot \left\{\sum_{k=1}^{n} |\beta_k|\right\}.$$

We emphasize that for different values of p the metric spaces l_n^p consist of the same objects, but that the distance between two fixed n-tuples varies, in general, with p. (Cf. Exercise 1.)

It is natural to consider the possibility of making n infinite in the preceding considerations. Everything goes through as before, provided, as might be expected, that we confine attention to those sequences $(\alpha_1, \alpha_2, \alpha_3, \ldots)$ which satisfy the condition $\sum_{k=1}^\infty |\alpha_k|^p < +\infty$. We sketch the essential ideas very briefly. If $\sum_{k=1}^\infty |\alpha_k|^p$ and $\sum_{k=1}^\infty |\beta_k|^p$ both converge, then by setting all the quantities γ_k appearing in (6–1) equal to zero we obtain

$$\left\{\sum_{k=1}^{n} |\alpha_k - \beta_k|^p\right\}^{1/p} \leqslant \left\{\sum_{k=1}^{n} |\alpha_k|^p\right\}^{1/p} + \left\{\sum_{k=1}^{n} |\beta_k|^p\right\}^{1/p}. \tag{6–3}$$

Fixing n temporarily on the left side and exploiting the assumed convergence of the infinite series mentioned previously, we obtain

$$\left\{\sum_{k=1}^{n} |\alpha_k - \beta_k|^p\right\}^{1/p} \leqslant \left\{\sum_{k=1}^{\infty} |\alpha_k|^p\right\}^{1/p} + \left\{\sum_{k=1}^{\infty} |\beta_k|^p\right\}^{1/p}.$$

Now we may let n approach $+\infty$ on the left, and we obtain

$$\left\{\sum_{k=1}^{\infty} |\alpha_k - \beta_k|^p\right\}^{1/p} \leqslant \left\{\sum_{k=1}^{\infty} |\alpha_k|^p\right\}^{1/p} + \left\{\sum_{k=1}^{\infty} |\beta_k|^p\right\}^{1/p}. \tag{6–4}$$

This shows that the series $\sum_{k=1}^\infty |\alpha_k - \beta_k|^p$ converges. Now if $\sum_{k=1}^\infty |\gamma_k|^p$ converges, we may replace α_k and β_k in the preceding inequality by

$\alpha_k - \gamma_k$ and $\gamma_k - \beta_k$, respectively, and we thus obtain

$$\left\{\sum_{k=1}^{\infty}|\alpha_k - \beta_k|^p\right\}^{1/p} \leqslant \left\{\sum_{k=1}^{\infty}|\alpha_k - \gamma_k|^p\right\}^{1/p} + \left\{\sum_{k=1}^{\infty}|\gamma_k - \beta_k|^p\right\}^{1/p}. \tag{6–5}$$

Thus, we have shown that the set of all sequences of complex numbers whose components are p-th power summable becomes a metric space if the distance between the sequences $(\alpha_1, \alpha_2, \alpha_3, \ldots)$ and $(\beta_1, \beta_2, \beta_3, \ldots)$ is defined by the left side of (6–5). This metric space is denoted l^p; in contrast to l_n^p, a sequence belonging to l^{p_1} need not belong to l^{p_2} if $p_1 \neq p_2$. (Cf. Exercise 2.)

Just as we obtained (6–3) as a generalization of (6–1), so we can generalize (6–2) to furnish the inequality

$$\sum_{k=1}^{\infty}|\alpha_k\beta_k| \leqslant \left\{\sum_{k=1}^{\infty}|\alpha_k|^p\right\}^{1/p} \cdot \left\{\sum_{k=1}^{\infty}|\beta_k|^q\right\}^{1/q}, \tag{6–6}$$

provided that the sequences $(\alpha_1, \alpha_2, \alpha_3, \ldots)$ and $(\beta_1, \beta_2, \beta_3, \ldots)$ belong to l^p and l^q, respectively. As before, we may define l^∞ as the collection of all bounded sequences, the distance between the sequences $(\alpha_1, \alpha_2, \alpha_3, \ldots)$ and $(\beta_1, \beta_2, \beta_3, \ldots)$ being defined as $\sup_{1 \leqslant k < \infty} |\alpha_k - \beta_k|$.

The inequalities (6–2) and (6–6), which are both easy consequences of the inequality developed in Theorem 2–2, are, like the latter inequality, also termed *Hölder's inequality;* similarly, (6–3) and (6–4) are termed *Minkowski's inequality.*

Finally, we consider briefly the analogue of the Riesz-Fischer theorem. As might be expected, the metric spaces l_n^p and l^p are complete for all p, including $p = +\infty$. Since the case of finite n is easily handled directly, or may easily be obtained as a corollary of the case of infinite n, we confine the statement of the following theorem to the l^p spaces.

Theorem 1: *For any p, $1 \leqslant p \leqslant +\infty$, the metric space l^p is complete. If the sequence $a_1, a_2, a_3, \ldots,$ where $a_k = (\alpha_1^{(k)}, \alpha_2^{(k)}, \alpha_3^{(k)}, \ldots)$, is Cauchy, then for each positive integer m the sequence $\alpha_m^{(1)}, \alpha_m^{(2)}, \alpha_m^{(3)}, \ldots$ converges to a number α_m; the sequence $(\alpha_1, \alpha_2, \alpha_3, \ldots)$ belongs to l^p, and is the limit of the given Cauchy sequence. (Less formally, we may say that the given sequence $a_1, a_2, a_3, \ldots$ converges component-wise to a member of l^p, say a, and $\lim_{k\to\infty} \rho_p(a, a_k) = 0$.)*

Proof: Corresponding to each vector a_k we associate the function $\tilde{a}_k$, defined on $[0, +\infty)$ as follows: $\tilde{a}_k(t) = \alpha_m^{(k)}$ for $t \in [m-1, m)$. It is obvious that $\tilde{a}_k \in L^p([0, +\infty))$ and that $\rho_p(a_j, a_k) = \|\tilde{a}_j - \tilde{a}_k\|_p$. Hence, the functions $\tilde{a}_1, \tilde{a}_2, \tilde{a}_3, \ldots$ form a Cauchy sequence in $L^p([0, +\infty))$. By the Riesz-Fischer theorem, this Cauchy sequence contains a subsequence which converges almost everywhere, and from the very simple form of the functions involved it is obvious that this subsequence must converge *everywhere*, not merely almost everywhere, on $[0, +\infty)$ to a function $\tilde{a}$ which is constant on each interval $[m - 1, m)$, $m = 1, 2, 3, \ldots$. If the value of $\tilde{a}(t)$ on $[m - 1, m)$ is denoted α_m, it is evident that the sequence* $a = (\alpha_1, \alpha_2, \alpha_3, \ldots)$ belongs to l^p (since $\tilde{a} \in L^p([0, +\infty))$) and that the sequence $a_1, a_2, a_3, \ldots$ converges to a.

The arguments presented in this section are, in fact, unnecessarily elaborate, for no reference is needed to integration. In Exercise 3 the reader is asked to provide direct proofs of the principal results of this section.

Exercises

1. Prove that the distance $\rho_p(a, b)$ between the n-tuples $a = (\alpha_1, \alpha_2, \ldots, \alpha_n)$ and $b = (\beta_1, \beta_2, \ldots, \beta_n)$ is independent of p iff at most one of the n quantities $\alpha_k - \beta_k$ is non-zero.
2. If $1 \leqslant p_1 < p_2 \leqslant +\infty$, show that $l^{p_1} \subset l^{p_2}$.
3. Prove the inequalities (6–2) and (6–3) and Theorem 1 without employing any of the theory of integration.

§7. SEPARABILITY OF L^p AND l^p

Definition 1: *A metric space is said to be separable if it possesses a dense subset consisting of a finite or countably infinite number of points.*

Examples

(*a*) Trivially, any metric space consisting of a finite or countably infinite number of points is separable, for the entire space is a dense subset of itself.

(*b*) The metric space defined in Example (*b*) of §1–2 is separable iff it consists of a finite or countably infinite number of points (for no *proper* subset of this space can be dense).

(*c*) In contrast to the preceding example, we note that R, although uncountable, is separable, for the countably infinite subset Q is dense. Similarly, the complex number system $\mathfrak{C}$ is dense, for the subset consisting of all *complex-rational numbers* (i.e., those whose real and imaginary parts are both rational) is both dense and countably infinite.

We now proceed to prove that both $L^p(A)$ and l^p are separable. (The separability of l_n^p, which should be obvious, is implicitly proven

* One must distinguish carefully between the objects $a, a_1, a_2, a_3, \ldots$, which are sequences of complex numbers (i.e., sequences in $\mathfrak{C}$) and the sequence $a_1, a_2, a_3, \ldots$ (which is a sequence in l^p).

during the proof for l^p.) We shall first consider l^p, for the argument is somewhat simpler in this case, and then we shall merely sketch the proof for $L^p(A)$.

Theorem 1: *l^p is separable (for $1 \leqslant p < +\infty$).*

PROOF: Let S be the subset of l^p consisting of those sequences containing only complex-rational terms of which only a finite number (if any) differ from zero. By an elementary argument (cf. Exercise 3) it is shown that S is countably infinite. Given any positive number ϵ and any member $a = (\alpha_1, \alpha_2, \alpha_3, \ldots)$ of l^p, we can, because of the convergence of the series $\sum_{k=1}^{\infty} |\alpha_k|^p$, choose an integer N so large that $\rho_p(a, a_N) < \epsilon/2$, where a_N is the sequence $(\alpha_1, \alpha_2, \alpha_3, \ldots, \alpha_N, 0, 0, 0, \ldots)$. Then it follows from the denseness of the complex-rational numbers in $\mathfrak{C}$ (and from the finiteness of N) that we can choose a vector

$$b = (\beta_1, \beta_2, \beta_3, \ldots, \beta_N, 0, 0, 0, \ldots)$$

belonging to S such that $\rho_p(a_N, b) < \epsilon/2$. It now follows by the triangle inequality that $\rho_p(a, b) < \epsilon$, and so it has been shown that S is dense, and hence, that l^p is separable. Note that this argument fails for $p = +\infty$; in fact, l^∞ is *not* separable. (Cf. Exercise 2.)

Theorem 2: *$L^p(A)$ is separable (for $1 \leqslant p < +\infty$).*

PROOF: We shall, for ease in exposition, confine attention to the case $p = 1$, $A = [0, 1]$, leaving it to the reader to convince himself that the argument can be modified easily to apply to any larger (finite) value of p and to any other measurable set A.

First we observe, by considering separately the real and imaginary parts of any member of $L^1(A)$, that it suffices to confine attention to real-valued functions; then, by expressing a real-valued function as the difference of its positive and negative parts, we see that it suffices to show that, given any non-negative real-valued function f belonging to $L^1(A)$ and any positive number ϵ, there exists a function g belonging to a fixed countable subset of $L^1(A)$ such that $\|f - g\|_1 < \epsilon$. If f is unbounded, we see by invoking the monotone convergence theorem that there exists a *bounded* non-negative member f_1 of $L^1(A)$ such that $\|f - f_1\|_1 < \epsilon/3$. (If f is bounded, we may merely choose $f_1 = f$.) By Lusin's theorem, there exists a *continuous* function f_2 such that $0 \leqslant f_2 \leqslant \sup f_1$ everywhere in A while the set $\{x \mid f_1(x) \neq f_2(x)\}$ has measure less than $\epsilon/(3 \sup f_1)$. It is then evident that $\|f_1 - f_2\|_1 < \epsilon/3$. Next, since f_2 is *uniformly* continuous, we can choose a real-valued step-function f_3 such that $|f_2(x) - f_3(x)| < \epsilon/3$ everywhere in A, so that $\|f_2 - f_3\|_1 = \int_0^1 |f_2 - f_3| < \epsilon/3$; furthermore, we may impose on f_3 the conditions that it assumes only rational values and that its discontinuities occur only at rational points of A. It then follows that $\|f - f_3\|_1 \leqslant \|f - f_1\|_1 + \|f_1 - f_2\|_1 + \|f_2 - f_3\|_1 < \epsilon$. Since the collection of step-functions satisfying the conditions imposed on f_3 is countable (the argument is virtually identical with that employed in proving that the set S employed in the proof of Theorem 1 is countable), the proof is complete.

An alternative to the last part of the proof is the following: By the Weierstrass approximation theorem (cf. Appendix D) and the denseness of Q in R, it is possible to choose for f_3 a polynomial with rational coefficients (instead of a step-function). Since such polynomials form a countable subset of $L^1(A)$, we conclude again that $L^1(A)$ is separable.

Taking account of the decomposition of each member of $L^1(A)$ into real and imaginary parts, we see that each of the following countable collections of functions is dense in $L^1(A)$:

(*a*) The collection of all step-functions having discontinuities only at rational points in A and assuming only complex-rational values.

(*b*) The collection of all polynomials possessing complex-rational coefficients.

In fact, each of these collections is also dense in $L^p(A)$ for $1 < p < +\infty$, and, with minor modifications in case (*a*), to any bounded (measurable) set A. These facts should become apparent when the reader extends the preceding proof to the case $p > 1$.

Exercises

1. Prove that any non-empty subset of a separable metric space is also separable.

2. Prove that l^∞ is not separable. Hint: Show, by imitating the diagonalization proof of the uncountability of R, that the subset of l^∞ consisting of sequences whose terms are exclusively zeros and ones is not separable.

3. Justify the assertion made in the proof of Theorem 1 that the set S is countable. Hint: Exploit repeatedly the fact that the cartesian product of two countable sets is also countable.

CHAPTER 4

NORMED LINEAR SPACES

In this chapter we shall present some of the most important ideas relating to normed linear spaces. The concept of a linear space involves, in its most general formulation, a field of scalars. However, we shall be interested exclusively in two fields of scalars, namely, the real and complex number systems, and, with occasional slight modifications (which will be pointed out at the appropriate places), the theory which we shall develop applies equally well to linear spaces over either of these two fields; in order to formulate our results in such a manner as to apply equally well to both cases, we shall usually employ the term "scalar" rather than "real number" or "complex number."

§1. LINEAR SPACES

The reader is, almost certainly, already acquainted, at least on an intuitive basis, with the concept of a linear space (also called "vector space"). We shall, therefore, present the basic ideas very concisely, trusting the reader to fill in the necessary details.

Definition 1: *A linear space is a non-empty collection of objects, called vectors, one of which is the zero vector (denoted o), which can be added pairwise and multiplied by any scalar in a manner consistent with the* *following laws*, which are to hold for all vectors f, g, h and all scalars α, β:*

(*a*) $f + g = g + f$, (*b*) $f + (g + h) = (f + g) + h$,
(*c*) $\alpha(f + g) = (\alpha f) + (\alpha g)$, (*d*) $\alpha(\beta f) = (\alpha\beta)f$,
(*e*) $(\alpha + \beta)f = (\alpha f) + (\beta f)$, (*f*) $1f = f$,
(*g*) $f + o = f$, (*h*) $0f = o$.

We shall refrain from proving, or even stating specifically, the vast number of immediate consequences of this definition. For example, we shall freely use an expression such as $e + f + g + h$, which, strictly speaking, becomes meaningful only when parentheses are suitably provided; since all (legitimate) distributions of parentheses fortunately lead to the same vector, we may thus speak unambiguously of the vector $e + f + g + h$. The proofs of this fact and innumerable others are virtually identical with those that the reader has encountered in studying the foundations of the real number system.

Examples

Many of the sets used in Chapter 1 to provide illustrations of metric spaces also provide illustrations of linear spaces. However, it must be kept clearly in mind that even though one and the same collection of objects may sometimes be thought of as *either* a metric space *or* a linear space, two essentially different ideas are involved: in the former case, we are concerned only with assigning a distance between any two members of the collection, while in the latter case we are concerned with adding members of the collection and multiplying them by scalars. As a matter of fact, we shall be particularly interested in collections of objects which are *both* metric spaces and linear spaces, but we must delay briefly the introduction of these mathematical entities.

(*a*) The most trivial example is furnished by the set consisting of a single object o satisfying the obvious rules: $o + o = o$, $\alpha o = o$ for any scalar α.

(*b*) The real number system, provided with the usual definition of addition and multiplication, constitutes a real linear space (i.e., a linear space over the real field).

(*c*) The complex number system, provided with the usual definition of addition and multiplication, constitutes a *real* linear space if we permit multiplication only by real numbers, while it constitutes a *complex* linear space if we permit multiplication by complex numbers.

(*d*) The class of all continuous real-valued functions defined on any interval (not necessarily bounded), if addition of two such functions and multiplication by any real scalar are defined in the obvious manner,

* For clarity in exposition we agree to denote vectors by lower case italic English letters and scalars by lower case Greek letters (except when specific numerical values appear as scalars).

becomes a real linear space. If, instead, we consider complex-valued functions, we have either a real or a complex linear space, according as we allow multiplication by real or by complex scalars.

(*e*) Similarly, we obtain linear spaces if in (*d*) we replace the condition of continuity by the condition of boundedness (but *not* if we impose the stronger restriction that $|f(x)| \leqslant 100$ for all points x in the specified interval [why?]).

(*f*) The set of ordered n-tuples of real numbers $(\alpha_1, \alpha_2, \ldots, \alpha_n)$ becomes a real linear space if we define addition of vectors and multiplication by real scalars in the obvious manner: $(\alpha_1, \alpha_2, \ldots, \alpha_n) + (\beta_1, \beta_2, \ldots, \beta_n) = (\alpha_1 + \beta_1, \alpha_2 + \beta_2, \ldots, \alpha_n + \beta_n)$ and $\gamma(\alpha_1, \alpha_2, \ldots, \alpha_n) = (\gamma\alpha_1, \gamma\alpha_2, \ldots, \gamma\alpha_n)$.

We conclude this introductory section with a few elementary definitions and theorems.

Definition 2: *A finite non-empty collection of vectors $\{f_1, f_2, \ldots f_n\}$ is said to be linearly independent if the equality $\alpha_1 f_1 + \alpha_2 f_2 + \cdots + \alpha_n f_n = o$ implies that $\alpha_1 = \alpha_2 = \cdots = \alpha_n = 0$. An infinite collection is said to be linearly independent if every finite (non-empty) subcollection is linearly independent.*

Definition 3: *A non-empty set S of vectors is termed a linear manifold if, for every pair of vectors $\{f, g\}$ belonging to S and every pair of scalars $\{\alpha, \beta\}$, the vector $\alpha f + \beta g$ also belongs to S.*

Theorem 1: *If the vectors $f_1, f_2, \ldots, f_n$ belong to a linear manifold S and if $\alpha_1, \alpha_2, \ldots, \alpha_n$ are arbitrary scalars, then $\alpha_1 f_1 + \alpha_2 f_2 + \cdots + \alpha_n f_n \in S$.*

PROOF: Trivial induction based on Definition 3.

Theorem 2: *Given any non-empty collection T of vectors, the set S of all vectors expressible as linear combinations of members of T (i.e., $\alpha_1 f_1 + \alpha_2 f_2 + \cdots + \alpha_n f_n$ for any choice of a finite set of vectors $f_1, f_2, \ldots, f_n$ belonging to T and any choice of the scalars $\alpha_1, \alpha_2, \ldots, \alpha_n$) is a linear manifold; furthermore, $T \subseteq S$, and if $\tilde{S}$ is any linear manifold such that $T \subseteq \tilde{S}$, then $S \subseteq \tilde{S}$; i.e., S is the smallest linear manifold containing T. The manifold S is said to be spanned by T, and is denoted $\mathfrak{M}(T)$.*

PROOF: Left to reader as Exercise 1.

Theorem 3: *If T is a linearly independent collection of vectors, then each member of $\mathfrak{M}(T)$ possesses a unique representation as a linear combination of members of T. (Note that this theorem holds whether the set T is finite or infinite.)*

PROOF: Left to reader as Exercise 2.

Definition 4: *A basis of a linear space L is a linearly independent collection T of vectors such that* $L = \mathfrak{M}(T)$. *A finite-dimensional linear space is one which is spanned by some finite collection of vectors; otherwise it is infinite-dimensional.*

Theorem 4: (*a*) *Every finite-dimensional linear space contains a basis.**

(*b*) *Every basis of a finite-dimensional linear space consists of a finite number of vectors. (This seems obvious, but the proof is a bit tricky.) Furthermore, any two bases of the same finite-dimensional linear space consist of the same number of vectors. This number, being independent of the particular choice of basis, thus represents a property of the given linear space, and is termed the dimension of the space; we employ the obvious notation* dim (*L*) *for the dimension of L.*

(*c*) *From every collection of vectors S which spans a finite-dimensional linear space L but does not constitute a basis of L, it is possible to extract a proper subset* $\tilde{S}$ *of S which does constitute a basis of L.*

(*d*) *Every linearly independent subset of a finite-dimensional linear space L which is not a basis can be enlarged to form a basis.*

PROOF: Left to reader as Exercise 3.

Theorem 5: *Any two linear spaces* L_1 *and* L_2 *(over the same field of scalars) of the same finite dimension are isomorphic; that is, it is possible to establish a one-to-one correspondence between the members of* L_1 *and* L_2 *which preserves the operations of addition of vectors and multiplication by scalars.*

PROOF: Let n be the common value of $\dim(L_1)$ and $\dim(L_2)$. According to Theorem 4, we can find a basis $\{f_1, f_2, \ldots, f_n\}$ of L_1 and a basis $\{g_1, g_2, \ldots, g_n\}$ of L_2. By Theorem 3, each member f of L_1 has a unique representation in terms of the f_k's:

$$f = \alpha_1 f_1 + \alpha_2 f_2 + \cdots + \alpha_n f_n.$$

Conversely, every choice of the scalars $\alpha_1, \alpha_2, \ldots, \alpha_n$ determines a unique member of L_1. The correspondence

$$\alpha_1 f_1 + \alpha_2 f_2 + \cdots + \alpha_n f_n \leftrightarrow \alpha_1 g_1 + \alpha_2 g_2 + \cdots + \alpha_n g_n$$

obviously provides an isomorphism between L_1 and L_2.

---

* This is also true of an infinite-dimensional space, but the proof requires transfinite induction or some equivalent logical principle. (Cf. Exercise *A*–6.)

Indeed, this argument shows that there is, for each positive integer n, essentially only one real and one complex n-dimensional linear space, namely, the set of all ordered n-tuples of scalars.

Exercises

1. Prove Theorem 2.
2. Prove Theorem 3.
3. Prove Theorem 4. (Not as trivial as it may appear! If this problem appears too difficult, the reader should consult a text on linear algebra.)

§2. NORMED LINEAR SPACES

The L^p-spaces which were considered at some length in Chapter 3 are linear spaces; this is clear from Theorems 1–1 and 2–3 of that chapter. However, these spaces possess an additional feature—with each member f of L^p is associated the real number $\|f\|_p$. We now proceed to discuss this additional feature in an abstract setting, and later we shall make numerous applications of it.

Definition 1: *A normed linear space is a linear space on which is defined a real-valued function, denoted* $\|\ \|$ *and termed a norm, satisfying the following properties:*

(*i*) $\|f\| = 0$ *iff* $f = o$, $\|f\| > 0$ *iff* $f \neq o$;

(*ii*) $\|\alpha f\| = |\alpha| \cdot \|f\|$ *for every scalar* α *and every vector* f (*homogeneity property of the norm*);

(*iii*) $\|f + g\| \leqslant \|f\| + \|g\|$ *for every pair of vectors* $\{f, g\}$.

Examples

(*a*) Theorems 1–1 and 2–4 of Chapter 3 show that L^p is a normed linear space.

(*b*) Either R or $\mathfrak{C}$, with the norm of any number defined as its absolute value.

(*c*) The linear space $C([a, b])$, with $\|f\| = \max_{a \leqslant x \leqslant b} |f(x)|$.

(*d*) The linear space $C([a, b])$, with $\|f\| = \int_a^b |f(x)|\, dx$.

(*e*) The linear space of all ordered n-tuples of scalars, with

$$\|(\alpha_1, \alpha_2, \ldots, \alpha_n)\| = \max_{1 \leqslant k \leqslant n} |\alpha_k|.$$

(*f*) The same linear space as in (*e*), but with $\|(\alpha_1, \alpha_2, \ldots, \alpha_n)\| = \{\sum_{k=1}^n |\alpha_k|^p\}^{1/p}$, where p is a finite constant $\geqslant 1$. (The validity of condition (*iii*) of Definition 1 is established in §3–6.)

The reader will certainly note the close relationship between some of the examples of metric spaces given in Chapter 1 on the one hand and some of the examples which we have just listed. The following theorem makes clear that every normed linear space is endowed in a very natural manner with the structure of a metric space. The elementary proof is left as Exercise 1.

Theorem 1: *A normed linear space becomes a metric space if the distance between two vectors is defined as follows:* $\rho(f, g) = \|f - g\|$. *(It is understood, of course, that* $-g$ *means* $(-1)g$ *and that* $f - g$ *means* $f + (-g)$.*)*

Whenever a normed linear space is treated as a metric space, it is understood that the distance function is the one *induced* by the norm in the manner explained in the preceding theorem.

Theorem 2: *For any vectors* f *and* g, $\|f \pm g\| \leqslant \|f\| + \|g\|$ *and* $\|f \pm g\| \geqslant |\, \|f\| - \|g\| \,|$.

PROOF: The first assertion, with the plus sign, is simply part (*iii*) of Definition 1. Replacing g by $-g$, we obtain $\|f - g\| \leqslant \|f\| + \|-g\|$, and by part (*ii*) of this same definition we obtain $\|-g\| = \|g\|$ (by choosing $\alpha = -1$).

Using what we have already established, we now prove the second part as follows: $\|f\| = \|(f \pm g) \mp g\| \leqslant \|f \pm g\| + \|g\|$, and hence $\|f \pm g\| \geqslant \|f\| - \|g\|$. Interchanging f and g and again exploiting part (*ii*) of Definition 1, we obtain $\|f \pm g\| \geqslant \|g\| - \|f\|$. Thus, $\|f \pm g\| \geqslant \max\{\|f\| - \|g\|, \|g\| - \|f\|\} = |\, \|f\| - \|g\| \,|$.

Part (*iii*) of Definition 1 together with the related inequalities established in the preceding theorem are often collectively termed the *triangle inequality*.

Theorem 3: *If the sequence* $f_1, f_2, f_3, \ldots$ *is Cauchy, the corresponding sequence of real numbers* $\|f_1\|, \|f_2\|, \|f_3\|, \ldots$ *is convergent (to a finite limit). If the first sequence converges, then* $\lim_{n\to\infty} \|f_n\| = \|\lim_{n\to\infty} f_n\|$.

PROOF: Given $\epsilon > 0$, there exists $N(\epsilon)$ such that whenever m and n exceed $N(\epsilon)$, $\|f_n - f_m\| < \epsilon$. By the latter part of Theorem 2,

$$|\, \|f_n\| - \|f_m\| \,| < \epsilon,$$

and so the sequence $\|f_1\|, \|f_2\|, \|f_3\|, \ldots$ is Cauchy; but since R is *complete*, the latter sequence actually converges.

If the original sequence converges to the vector f, we obtain from the latter part of Theorem 2 the inequality $\big| \|f\| - \|f_n\| \big| \leqslant \|f - f_n\|$. Letting $n \to \infty$, we see that $\|f\| - \|f_n\| \to 0$, or $\|f_n\| \to \|f\|$.

Although all the necessary tools are now available, we shall dispense with the needless task of formally stating and proving a whole host of theorems which are obviously extensions of theorems concerning sequences of real numbers, such as that the convergence of the sequence $f_1, f_2, f_3, \ldots$ to f and the convergence of $g_1, g_2, g_3, \ldots$ to g imply the convergence of the sequence $f_1 + g_1, f_2 + g_2, f_3 + g_3, \ldots$ to $f + g$.

Theorem 4: *Let N be a normed linear space (not necessarily finite-dimensional) and let $f_1, f_2, \ldots, f_n$ be any finite linearly independent collection of vectors. Then there exists a positive number δ such that, for every choice of the scalars $\alpha_1, \alpha_2, \ldots, \alpha_n$, the inequality*

$$\|\alpha_1 f_1 + \alpha_2 f_2 + \cdots + \alpha_n f_n\| \geqslant \delta(|\alpha_1| + |\alpha_2| + \cdots + |\alpha_n|)$$

holds. (Roughly speaking, it is impossible for large scalars to combine so as to furnish a small vector.)

PROOF: By the homogeneity property of the norm (cf. Definition 1), it suffices to show that $\|\alpha_1 f_1 + \alpha_2 f_2 + \cdots + \alpha_n f_n\|$ cannot come arbitrarily close to zero when the α's satisfy the restriction $|\alpha_1| + |\alpha_2| + \cdots + |\alpha_n| = 1$. If the theorem were false, we could choose a sequence of vectors $g_1, g_2, g_3, \ldots$, each expressible in the form

$$g_j = \alpha_{1,j} f_1 + \alpha_{2,j} f_2 + \cdots + \alpha_{n,j} f_n, \qquad \sum_{k=1}^{n} |\alpha_{k,j}| = 1$$

such that $\|g_j\| \to 0$ as $j \to \infty$. Since each of the numbers $|\alpha_{k,j}|$ is certainly $\leqslant 1$, we can (by the Bolzano-Weierstrass theorem) select a subsequence of the g_j's for which the corresponding subsequence of the numbers $\alpha_{1,j}$ converges to a number α_1; then we can extract from this subsequence of the g_j's a further subsequence such that the corresponding subsequence of the numbers $\alpha_{2,j}$ converges to a number α_2; by a finite number of such steps (since n is finite!) we obtain a subsequence of the g_j's, say $h_1, h_2, h_3, \ldots$, where

$$h_i = \beta_{1,i} f_1 + \beta_{2,i} f_2 + \cdots + \beta_{n,i} f_n, \qquad \sum_{k=1}^{n} |\beta_{k,i}| = 1$$

and $\lim_{i \to \infty} \beta_{k,i} = \alpha_k$. Then clearly $h_i \to h$, where

$$h = \alpha_1 f_1 + \alpha_2 f_2 + \cdots + \alpha_n f_n, \qquad \sum_{k=1}^{n} |\alpha_k| = 1.$$

Since the f's are linearly independent and the α's do not all vanish, $h \neq o$; on the other hand, clearly $\|h\| = \lim_{j\to\infty} \|g_j\| = 0$, and so $h = o$. This contradiction proves the existence of the positive number δ referred to in the statement of the theorem.

Corollary: *Let two norms, $\| \ \|$ and $\| \ \|'$, be defined on a finite-dimensional linear space. Then there exist positive numbers ϵ_1 and ϵ_2 such that, for every vector f, the inequalities $\epsilon_1 \|f\| \leqslant \|f\|' \leqslant \epsilon_2 \|f\|$ hold. A subset S which is open with respect to one of these norms is also open with respect to the other norm. (Thus, all norms are topologically equivalent, in the sense that they determine the same collection of open subsets.)*

PROOF: Select a basis $\{f_1, f_2, \ldots, f_n\}$ of the space, and let any vector f be expressed in the unique form $f = \alpha_1 f_1 + \alpha_2 f_2 + \cdots + \alpha_n f_n$. By Theorem 4, $\|f\| \geqslant \delta(|\alpha_1| + |\alpha_2| + \cdots + |\alpha_n|)$; on the other hand, by the triangle inequality we obtain $\|f\|' \leqslant m(|\alpha_1| + |\alpha_2| + \cdots + |\alpha_n|)$, where $m = \max\{\|f_1\|', \|f_2\|', \ldots, \|f_n\|'\}$. Hence, $\|f\|' \leqslant \frac{m}{\delta}\|f\| = \epsilon_2 \|f\|$. The other inequality is now obtained by interchanging the roles of $\| \ \|$ and $\| \ \|'$ in the preceding argument. Now, if S is open with respect to the norm $\| \ \|$ and if $x \in S$, then there exists a positive number r such that $S_r(x) \subseteq S$, where the norm $\| \ \|$ is employed in defining the open ball $S_r(x)$; from the previous results it is evident that, for sufficiently small $\tilde{r}$, $S'_{\tilde{r}}(x) \subseteq S$, where the norm $\| \ \|'$ is now employed. Hence, S is also open with respect to the norm $\| \ \|'$. We emphasize that this corollary fails in infinite-dimensional linear spaces (cf. Exercise 5.)

Theorem 5: *Every finite-dimensional normed linear space is complete.*

PROOF: Choose any basis $f_1, f_2, \ldots, f_n$ and any Cauchy sequence $g_1, g_2, \ldots$ and express each g_j in the (uniquely determined) form

$$g_j = \alpha_{1,j} f_1 + \alpha_{2,j} f_2 + \cdots + \alpha_{n,j} f_n.$$

Given $\epsilon > 0$, there exists $N(\epsilon)$ such that whenever i and j exceed $N(\epsilon)$, $\epsilon > \|g_j - g_i\| = \|\sum_{k=1}^n (\alpha_{k,j} - \alpha_{k,i}) f_k\| \geqslant \delta \sum_{k=1}^n |\alpha_{k,j} - \alpha_{k,i}|$. (The last step is justified by Theorem 4.) Hence, each of the n sequences of scalars $\alpha_{k,1}, \alpha_{k,2}, \alpha_{k,3}, \ldots$ is Cauchy and therefore convergent to limits α_k, $1 \leqslant k \leqslant n$. It then follows readily from the triangle inequality that $\|g - g_j\| \to 0$, where $g = \alpha_1 f_1 + \alpha_2 f_2 + \cdots + \alpha_n f_n$. Thus, any Cauchy sequence is, in fact, convergent, and so the normed linear space is complete.

Theorem 6: *Every finite-dimensional normed linear space is locally compact; i.e., from every bounded sequence of vectors $g_1, g_2, g_3, \ldots$ it is* *possible to choose a convergent subsequence.* (*Cf. Definition* 10–3 *and Theorem* 10–1 *of Chapter* 1.)

We leave the proof to the reader as Exercise 2; it is very closely related to the proof of the preceding theorem. It should be noticed that Theorem 6 is a straightforward extension of the Bolzano-Weierstrass theorem; indeed, it reduces to the latter when the normed linear space under consideration is one-dimensional.

Corollary: *Any finite-dimensional linear manifold of a normed linear space is closed.*

PROOF: Any limit-point f of the manifold is expressible as the limit of a sequence of distinct vectors contained in the manifold. Since a convergent sequence is Cauchy and since the manifold is complete (by Theorem 5), the limit-point f must belong to the manifold, and the latter is therefore closed.

Definition 2: *A subspace of a normed linear space N is a closed linear manifold of N.*

Theorem 7: *Let N be a normed linear space, let S be a subspace which is a proper subset of N, and let α be any positive number <1. Then there exists a vector in N whose norm is* 1 *and whose distance from S exceeds α.*

PROOF: Let f be any vector in $N - S$; since S is closed, the distance from f to S is some positive number δ. Hence there exists a vector g in S such that $\delta < \|f - g\| < \delta/\alpha$. Now $f - g \notin S$ (why?), and $\|f - g\| = \|(f - g) - o\|$; since $o \in S$, the distance between $f - g$ and S is $<\delta/\alpha$ and $\geqslant \delta$. Hence, $(f - g)/\|f - g\|$ is a vector of norm 1 whose distance from S is $\geqslant \delta/\|f - g\| > \delta/(\delta/\alpha) = \alpha$.

Definition 3: *A normed linear space which is complete* (*with respect to the metric induced by the norm*) *is termed a Banach space.*

Remark: Note that the first part of the Riesz-Fischer theorem simply asserts that each L^p is a Banach space, while Theorem 6–1 of Chapter 3 asserts the same for each l^p.

Theorem 8: *The metric in any normed linear space is translation-invariant; by this we mean that if f, g, and h are any vectors, then $\rho(f, g) = \rho(f + h, g + h)$.*

PROOF: $\rho(f, g) = \|f - g\| = \|(f + h) - (g + h)\| = \rho(f + h, g + h)$.

Definition 4: *A (non-empty) subset S of a linear space (not necessarily normed) is said to be convex if for every pair of vectors* $\{f, g\}$ *belonging to S and for every real number* α *in the interval* $[0, 1]$ *the vector* $\alpha f + (1 - \alpha)g$ *also belongs to S. (Note that the scalar* α *is confined to real values, even if the linear space is over the complex field.)*

Theorem 9: *Any ball, open or closed, in a normed linear space is convex.*

PROOF: By the preceding theorem, it suffices to consider a ball centered at the vector o. The vectors f and g belong to the closed ball $S_r[o]$ iff $\|f\| \leqslant r$, $\|g\| \leqslant r$. Then, for any number α in the interval $[0, 1]$ we obtain $\|\alpha f + (1 - \alpha)g\| \leqslant \|\alpha f\| + \|(1 - \alpha)g\| = \alpha \|f\| + (1 - \alpha)\|g\| \leqslant (\alpha + [1 - \alpha])r = r$, and so $\alpha f + (1 - \alpha)g$ also belongs to $S_r[o]$, which is thus shown to be convex. A similar proof holds for the open ball $S_r(o)$.

As indicated in Definition 4, the concept of convexity is independent of norm; it depends only on the linear space under consideration. On the other hand, a particular ball, $S_r(x)$ or $S_r[x]$, will depend not only on r and x, but also on the particular norm that is defined on the space; however, according to Theorem 9, the ball will be convex for any choice of the norm.

It is instructive to consider in detail a very simple particular example. Let L be the two-dimensional real linear space consisting of all ordered pairs of real numbers, with the usual definition of addition of vectors and multiplication by scalars. We can visualize L as the familiar plane of analytic geometry, but to begin with only the vector-space operations are defined; no concept of distance is involved at first. Then for each choice of the number p, $1 \leqslant p \leqslant +\infty$, let us norm L by defining $\|a\|_p$, where $a = (\alpha_1, \alpha_2)$, to be $(|\alpha_1|^p + |\alpha_2|^p)^{1/p}$ (with the understanding that for $p = +\infty$ the norm of a is taken to be $\max\{|\alpha_1|, |\alpha_2|\}$). The accompanying Figure 2 shows the open and closed unit balls, $S_1(o)$ and $S_1[o]$, for $p = 1$, $p = 2$, and $p = +\infty$, the open ball consisting in each case of the region enclosed by the correspondingly labeled curve, while the closed ball consists of the open ball together with the curve. For $p = 1$ and $p = +\infty$ the curve is a square, while for $p = 2$ the curve is a circle; in each case we observe that the ball (open or closed) is convex, in agreement with Theorem 9. It is easily seen that, as p increases from 1 to $+\infty$, the unit ball (open or closed) grows steadily. (One could show by the familiar methods of elementary calculus that the ball is convex, but Theorem 9, together with the results of §3–6, assure this result.)

It is instructive to observe what happens if we retain the preceding definition of $\|a\|_p$, but allow p to assume values in the open interval $(0, 1)$. In particular, for $p = \frac{1}{2}$ we obtain the region bounded by the innermost curve in Figure 2; this region is obviously not convex. A similar result is

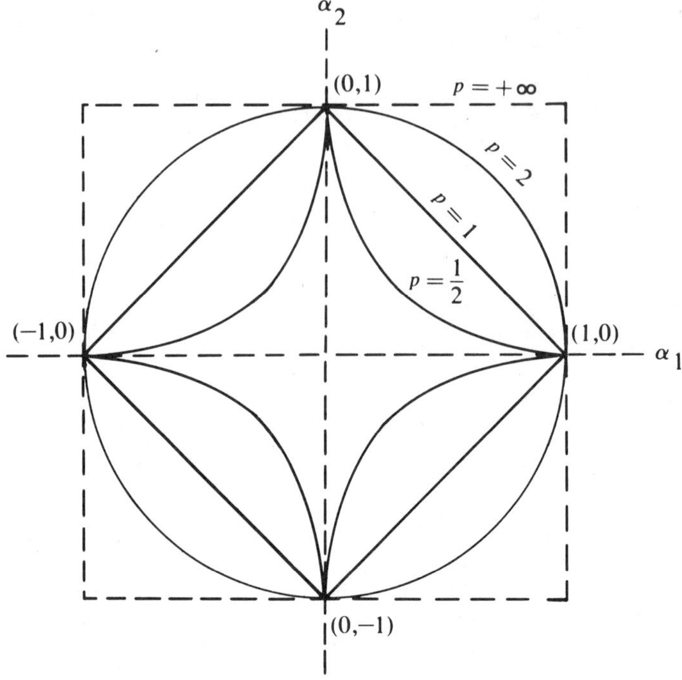

Figure 2.

obtained for any other choice of p in (0, 1), and so it follows that the restriction $p \geqslant 1$ is a necessary, as well as sufficient, condition that the defined "p-norm" should actually be a norm. (Cf. Exercise 3.)

Exercises

1. Prove Theorem 1.

2. Prove Theorem 6.

3. Find two ordered pairs of real numbers, $a = (\alpha_1, \alpha_2)$ and $b = (\beta_1, \beta_2)$, such that $\|a + b\|_p > \|a\|_p + \|b\|_p$ for every p satisfying $0 < p < 1$.

4. Show that for any p satisfying the inequalities $1 \leqslant p \leqslant +\infty$ the subset of l^p consisting of sequences containing only a finite number of non-zero entries is a normed linear space but not a Banach space.

5. Show by means of a specific example that the corollary to Theorem 4 fails in infinite-dimensional linear spaces. (Hint: Make use of Exercise 2–3 of Chapter 1.)

§3. INNER-PRODUCT SPACES

The reader will recall from the elements of euclidean vector analysis the formula

$$\cos\theta = \frac{\alpha_1\beta_1 + \alpha_2\beta_2 + \alpha_3\beta_3}{\|a\| \cdot \|b\|},$$

where θ denotes the angle between the non-zero vectors a and b whose cartesian components are α_1, α_2, α_3 and β_1, β_2, β_3, respectively; of course, $\|a\|$ and $\|b\|$ are the lengths of the vectors a and b, respectively. The definition of a normed linear space does not permit the introduction, in any natural manner, of the concept of the angle between two vectors. We now introduce the inner-product spaces, which admit, as will be seen, the concept of angle between any two non-zero vectors, at least in the case when the *real* field of scalars is employed. Actually, the concept of angle is itself of rather minor significance; it is the concept of orthogonality, or perpendicularity, which is really significant, and this can be defined whether the real or complex scalars are employed.

Definition 1: *A complex* inner-product space is a complex linear space L together with a complex-valued function, denoted (,) and called the inner product, defined on $L \times L$ and having the following properties:*

(*i*) $(f, f) > 0$ *if* $f \neq o$, $(o, o) = 0$;

(*ii*) $(f, g) = \overline{(g, f)}$ *for all vectors* f, g;

(*iii*) $(f + g, h) = (f, h) + (g, h)$ *for all vectors* f, g, h;

(*iv*) $(\alpha f, g) = \alpha(f, g)$ *for all vectors* f, g *and scalars* α.

In Exercise 1 we list a number of simple but important consequences of this definition, which will be used frequently without comment.

Examples

(*a*) The set of all ordered n-tuples of scalars† with $(a, b) = \alpha_1\bar{\beta}_1 + \alpha_2\bar{\beta}_2 + \cdots + \alpha_n\bar{\beta}_n$, where $a = (\alpha_1, \alpha_2, \ldots, \alpha_n)$ and $b = (\beta_1, \beta_2, \ldots, \beta_n)$.

(*b*) The set of all continuous complex-valued functions defined on a compact interval $[a, b]$, with $(f, g) = \int_a^b f\bar{g}$.

(*c*) $L^2(A)$, with (f, g) defined as in (*b*). (The Hölder inequality shows that $\int f\bar{g}$ makes sense when f and g belong to $L^2(A)$.) As in previous

---

* For convenience we specifically formulate our definition for complex spaces; with a few changes, almost all quite obvious, the remaining material of this chapter applies equally well to real spaces. A bar over a complex number denotes the conjugate of that number, so no confusion with the previous use of the bar to signify closure of a set can arise.

† We refrain henceforth from stating the obvious definitions of addition and multiplication by scalars.

discussions involving integration, A denotes any subset of R possessing positive measure ($+\infty$ permitted).

Theorem 1 (Schwarz Inequality): *For any two vectors f and g, $|(f, g)|^2 \leqslant (f, f)(g, g)$, equality holding iff f and g are linearly dependent.*

PROOF: The assertion is obvious if either f or g is the vector o, for then both sides of the inequality reduce to zero. We may therefore assume that $f \neq o$, $g \neq o$. Consider $(f - \lambda g, f - \lambda g)$ as a function of the *real* variable λ. By (i) of Definition 1, this expression is always non-negative; since $(f - \lambda g, f - \lambda g) = (f, f) - 2\lambda \operatorname{Re}(f, g) + \lambda^2(g, g)$, we conclude that the discriminant of the quadratic function appearing on the right is non-positive (and real). Thus, $\{\operatorname{Re}(f, g)\}^2 \leqslant (f, f)(g, g)$. If $(f, g) \geqslant 0$, then $\operatorname{Re}(f, g) = |(f, g)|$, and we have the desired result. Otherwise, we can choose the scalar α such that $|\alpha| = 1$ and $\alpha(f, g) \geqslant 0$; replacing f in the preceding argument by αf, we obtain $\{\operatorname{Re}(\alpha f, g)\}^2 \leqslant (\alpha f, \alpha f)(g, g) = \alpha\bar{\alpha}(f, f)(g, g) = (f, f)(g, g)$, and hence $|(\alpha f, g)|^2 \leqslant (f, f)(g, g)$. Since $|(\alpha f, g)|^2 = |\alpha|^2 \, |(f, g)|^2 = |(f, g)|^2$, we again obtain $|(f, g)|^2 \leqslant (f, f)(g, g)$.

If we trace through the steps of the argument, we easily find that equality holds iff $f - \lambda g = o$ for *some* choice of λ (not necessarily real, because of the use of the scalar α in the later stages of the argument). Taking account of this observation and recalling the trivial cases that were set aside at the beginning of the proof, we obtain the final assertion of the theorem.

We give a second proof, which the reader may find instructive: As before we may assume that $g \neq o$, so that $(g, g) > 0$. We may then form the vector $h = f - ((f, g)/(g, g))g$. A simple calculation furnishes the equality $(h, h) = (f, f) - |(f, g)|^2/(g, g)$; since $(h, h) \geqslant 0$, we conclude that $|(f, g)|^2 \leqslant (f, f)(g, g)$, equality holding iff $h = o$, or $f = \lambda g$, where $\lambda = (f, g)/(g, g)$.

Theorem 2: *For any two vectors f and g, $(f + g, f + g)^{1/2} \leqslant (f, f)^{1/2} + (g, g)^{1/2}$, equality holding iff f and g are positive multiples of each other (excluding the trivial case that either f or g is the vector o).*

PROOF: Using Theorem 1, we obtain the following chain of equalities and inequalities:

$$\begin{aligned}
(f + g, f + g) &= (f, f) + (f, g) + (g, f) + (g, g) \\
&= (f, f) + 2 \operatorname{Re}(f, g) + (g, g) \\
&\leqslant (f, f) + 2\,|(f, g)| + (g, g) \\
&\leqslant (f, f) + 2(f, f)^{1/2} \cdot (g, g)^{1/2} + (g, g) \\
&= [(f, f)^{1/2} + (g, g)^{1/2}]^2.
\end{aligned}$$

Taking the square root of the initial and final expressions, we obtain the desired result. Equality is seen to hold iff $\text{Re}\,(f, g) = |(f, g)| = (f, f)^{1/2}(g, g)^{1/2}$, and these two equalities lead to the condition stated in the theorem.

It is now clear that $(f, f)^{1/2}$ satisfies all the conditions imposed upon a norm, and so we use the notation $\|f\|$. Thus, every inner-product space becomes a normed linear space, and hence a metric space, with the understanding that $\|f\| = (f, f)^{1/2}$ and $\rho(f, g) = \|f - g\|$. We note that in a *real* inner-product space the Schwarz inequality assumes the form $-1 \leqslant (f, g)/(\|f\| \cdot \|g\|) \leqslant 1$, so that, exactly as in euclidean vector analysis, we can define the angle between any two non-zero vectors, as asserted in the first paragraph of this section.

Theorem 3: (*a*) *For any vectors f and g,* $\|f + g\|^2 + \|f - g\|^2 = 2\,\|f\|^2 + 2\,\|g\|^2$.

(*b*) *For any vectors f and g,* $(f, g) = \frac{1}{4}\{\|f + g\|^2 + i\,\|f + ig\|^2 - \|f - g\|^2 - i\,\|f - ig\|^2\}$.

PROOF: Left as Exercise 2.

Theorem 4: *The inner product* $(\alpha f, g)$ *depends continuously on the scalar* α *and the vectors* f, g*—that is, given* $\epsilon > 0$ *and* α, f, g*, there exists a positive number* δ *(depending on* $\epsilon, \alpha, f,$ *and* g*) such that* $|(\beta\tilde{f}, \tilde{g}) - (\alpha f, g)| < \epsilon$ *whenever* $|\beta - \alpha| < \delta$, $\|f - \tilde{f}\| < \delta$, $\|g - \tilde{g}\| < \delta$.

PROOF: Left as Exercise 3.

Definition 2: *The vectors f and g are said to be orthogonal if* $(f, g) = 0$. *(Clearly* $(f, g) = 0$ *iff* $(g, f) = 0$*, so that orthogonality is a symmetric relation. Also, o is the only vector which is orthogonal to itself.)*

Theorem 5: *Let A be any non-empty collection of vectors. Then the set of all vectors orthogonal to every vector in A is a subspace, called the orthogonal complement of A and denoted* $A^{\perp}$.

PROOF: Suppose g and h are both orthogonal to f. Then

$$(f, \alpha g + \beta h) = \bar{\alpha}(f, g) + \bar{\beta}(f, h) = 0 + 0 = 0;$$

hence $\alpha g + \beta h$ is orthogonal to f, and so $A^{\perp}$ is certainly a linear manifold. If g is a limit-point of $A^{\perp}$, we can find a sequence $g_1, g_2, \ldots$ in $A^{\perp}$ which converges to g. For every vector f in A, $(f, g) = (f, g_n) + (f, g - g_n) = (f, g - g_n)$, since $(f, g_n) = 0$. Thus, $|(f, g)| = |(f, g - g_n)| \leqslant \|f\| \cdot \|g - g_n\|$. Letting $n \to \infty$ and observing that (f, g) is independent of n, we conclude that $(f, g) = 0$, so that $g \in A^{\perp}$. Thus $A^{\perp}$ is closed, and is therefore a subspace.

Definition 3: *A collection of vectors is said to be orthogonal if every two distinct members of the collection are orthogonal. An orthogonal collection of vectors each having unit norm is called orthonormal. (A vector having unit norm is often termed a unit vector.)*

Theorem 6: *Any orthonormal collection of vectors is linearly independent.*

PROOF: If $f_1, f_2, \ldots, f_n$ constitute any finite set of vectors from this collection and if $\alpha_1 f_1 + \alpha_2 f_2 + \cdots + \alpha_n f_n = o$, then

$$\begin{aligned} 0 &= \|\alpha_1 f_1 + \alpha_2 f_2 + \cdots + \alpha_n f_n\|^2 \\ &= (\alpha_1 f_1 + \alpha_2 f_2 + \cdots + \alpha_n f_n, \alpha_1 f_1 + \alpha_2 f_2 + \cdots + \alpha_n f_n) = \sum_{k=1}^{n} |\alpha_k|^2. \end{aligned}$$

(The last equality is based directly on the orthonormality of the f's.) Thus, $\alpha_1 = \alpha_2 = \cdots = \alpha_n = 0$, and so the f's must be linearly independent.

Theorem 7: *If $f_1, f_2, \ldots, f_n$ are orthonormal and if g is expressible as a linear combination of the f's, then the combination must be $(g, f_1)f_1 + (g, f_2)f_2 + \cdots + (g, f_n)f_n$.*

PROOF: By assumption, there exist scalars $\alpha_1, \alpha_2, \ldots, \alpha_n$ such that $g = \alpha_1 f_1 + \alpha_2 f_2 + \cdots + \alpha_n f_n$. For any index k, $1 \leqslant k \leqslant n$, we obtain $(g, f_k) = (\alpha_1 f_1 + \alpha_2 f_2 + \cdots + \alpha_n f_n, f_k) = \alpha_k (f_k, f_k) + \sum_{j \neq k} \alpha_j (f_j, f_k) = \alpha_k$, since $(f_j, f_k) = 0$ if $j \neq k$.

Theorem 8: *Let $f_1, f_2, \ldots f_n$ be any linearly independent finite collection of vectors. Then there exists an orthonormal collection $h_1, h_2, \ldots, h_n$ which spans the same linear manifold as that spanned by the f's.*

PROOF: We merely sketch the principal idea, leaving the detailed proof as Exercise 4. Let $g_1 = f_1$, $g_2 = f_2 + \alpha_{12} f_1$, $g_3 = f_3 + \alpha_{23} f_2 + \alpha_{13} f_1$, and so forth, where the α's are so chosen that the g's form a collection of orthogonal non-zero vectors. (The fact that such a choice of α's is possible [and in a unique manner] rests on the linear independence of the f's and is the essential point in a rigorous proof.) Then, clearly, any vector expressible as a linear combination of the f's is also expressible as a linear combination of g's, and conversely. By setting $h_k = (1/\|g_k\|)g_k$ we obtain the desired orthonormal collection.

The procedure just described for obtaining an orthonormal collection of vectors spanning the same linear manifold as that spanned by a given (finite) collection of vectors is termed the *Gram-Schmidt process*. It obviously extends to a countably infinite linearly independent collection of vectors.

Theorem 9: *Let $h_1, h_2, \ldots, h_n$ be a finite orthonormal collection of vectors and let f be any vector. Then, among all choices of the scalars $\alpha_1, \alpha_2, \ldots, \alpha_n$, the (non-negative real) quantity*

$$\|f - (\alpha_1 h_1 + \alpha_2 h_2 + \cdots \alpha_n h_n)\|^2$$

is minimized by the unique choice $\alpha_k = (f, h_k)$, $1 \leqslant k \leqslant n$, and the minimum of the preceding expression is $\|f\|^2 - \sum_{k=1}^n |(f, h_k)|^2$.

PROOF: Let $\alpha_k = (f, h_k) + \beta_k$. An elementary computation (which must, of course, take account of the orthonormality of the h's), furnishes the equality $\|f - (\alpha_1 h_1 + \alpha_2 h_2 + \cdots + \alpha_n h_n)\|^2 = \|f\|^2 - \sum_{k=1}^n |(f, h_k)|^2 + \sum_{k=1}^n |\beta_k|^2$. The right side is clearly minimized by choosing $\beta_1 = \beta_2 = \cdots = \beta_n = 0$, and any other choice will increase the right side. Hence, the theorem is proved.

Corollary: *Let $\mathfrak{M}$ be any finite-dimensional linear subspace in an inner-product space, and let f be any vector in the space. Then there exists a unique vector $\tilde{g}$ in $\mathfrak{M}$ which minimizes $\|f - g\|$ among all vectors g in $\mathfrak{M}$; the vector $\tilde{g}$ is such that $f - \tilde{g} \in \mathfrak{M}^\perp$.*

PROOF: Construct an orthonormal basis $h_1, h_2, \ldots, h_n$ in $\mathfrak{M}$, apply Theorem 9, and observe that the vector

$$f - \{(f, h_1)h_1 + (f, h_2)h_2 + \cdots + (f, h_n)h_n\}$$

is orthogonal to each h_k, $1 \leqslant k \leqslant n$. (The reader will find it helpful to look upon this corollary as a generalization of the euclidean theorem which asserts that the shortest distance from a point to a line is furnished by the perpendicular.)

Corollary (Bessel's Inequality): *Let $h_1, h_2, \ldots$ be any orthonormal collection of vectors, finite or countably infinite. Then for any vector f the inequality $\sum |(f, h_k)|^2 \leqslant \|f\|^2$ holds. (If the h's constitute a countably infinite collection, the convergence of the infinite series $\sum |(f, h_k)|^2$ (to a finite sum) is part of the conclusion, not an added hypothesis.)*

PROOF: Referring to Theorem 9 and exploiting the fact that $\|f - (\alpha_1 h_1 + \alpha_2 h_2 + \cdots + \alpha_n h_n)\|^2 \geqslant 0$ for all choices of the α's, we conclude that $\sum_{k=1}^n |(f, h_k)|^2 \leqslant \|f\|^2$ for every positive integer n. Letting

n increase without bound, we conclude that the series $\sum_{k=1}^{\infty} |(f, h_k)|^2$ converges, and that the sum is $\leqslant \|f\|^2$, when the collection of h's is countably infinite. (Cf. Exercise 6.)

This corollary demonstrates, in particular, that the limiting relation $\lim_{n\to\infty} (f, h_n) = 0$ must hold when the h's constitute a countably infinite collection.

Exercises

1. (*a*) Prove that $(o, h) = (h, o) = 0$ for every vector h, so that the condition $(o, o) = 0$ appearing in Definition 1 is superfluous.
   (*b*) Prove that $(f, \alpha g) = \bar{\alpha}(f, g)$ for all vectors f, g and scalars α.
   (*c*) Prove that $(f, g + h) = (f, g) + (f, h)$ for all vectors f, g, h.
2. Prove Theorem 3. Why is part (*a*) called the parallelogram law?
3. Prove Theorem 4.
4. Work out the detailed proof of Theorem 8.
5. Prove that the finite collection $\{f_1, f_2, \ldots, f_n\}$ is linearly independent iff the expression
$$\sum_{j,k=1}^{n} (f_j, f_k)\alpha_j\bar{\alpha}_k$$
   is positive definite—i.e., positive for all choices of the scalars $\alpha_1, \alpha_2, \ldots, \alpha_n$ except the trivial one $\alpha_1 = \alpha_2 = \cdots = \alpha_n = 0$.
6. Suppose that the collection $\{h_i\}_{i\in I}$ is orthonormal and that the index-set I is uncountably infinite. Prove that, given any vector f, the inequality $(f, h_i) \neq 0$ holds only for a finite or countably infinite set of indices $i \in I$ and that $\sum |(f, h_i)|^2$ converges to a sum $\leqslant \|f\|^2$, where the summation is performed only over those indices for which $(f, h_i) \neq 0$.
7. Give an example of an inner-product space which contains an uncountably infinite orthonormal collection of vectors.

§4. HILBERT SPACES

Definition 1: *A Hilbert space is an inner-product space which is complete (in the norm induced by the inner product). (Frequently this term is restricted to infinite-dimensional spaces, and sometimes the additional restriction of separability is imposed.)*

Examples

(*a*) The Riesz-Fischer theorem guarantees the completeness of $L^2(A)$, the inner product being that defined in Example (*c*) following Definition 3–1.

(*b*) The results of §3–6 and §3–7 show that l^2 becomes a separable infinite-dimensional Hilbert space when the inner product of the vectors $a = (\alpha_1, \alpha_2, \alpha_3, \ldots)$ and $b = (\beta_1, \beta_2, \beta_3, \ldots)$ is defined as $\sum_{k=1}^{\infty} \alpha_k \bar{\beta}_k$. Henceforth, it is always understood that l^2 is a Hilbert space with the inner product as just defined.

(*c*) From Exercise 2–4 it is apparent that the subset of l^2 consisting of all sequences containing only a finite number of non-zero terms is an infinite-dimensional incomplete inner-product space.

Theorem I: *Let S be a non-empty closed convex subset of a Hilbert space H and let f be any vector in H. Then there exists a vector g in S such that $\|f - g\| < \|f - h\|$ whenever h is any member of S other than g. (From the very statement of the theorem it is clear that g is unique.)*

PROOF: By translation-invariance (cf. Theorem 2–8) of the metric and the trivial fact that convexity is preserved under translation, we may confine attention to the particular case that $f = o$, so that the theorem may be restated in the simpler form: *Any non-empty closed convex subset of a Hilbert space contains a unique vector of smallest norm.* Let δ denote the distance from the set S to the vector o (i.e., $\delta = \inf_{g \in S} \|g\|$). Then we can choose a sequence $g_1, g_2, g_3, \ldots$ of members of S such that $\|g_n\| \to \delta$. By the parallelogram law,

$$\left\| \frac{g_n - g_m}{2} \right\|^2 + \left\| \frac{g_n + g_m}{2} \right\|^2 = \tfrac{1}{2} \|g_n\|^2 + \tfrac{1}{2} \|g_m\|^2.$$

Since $\frac{1}{2}(g_n + g_m)$ $(= \frac{1}{2}g_n + (1 - \frac{1}{2})g_m) \in S$, the second term on the left is $\geqslant \delta^2$. Hence,

$$\left\| \frac{g_n - g_m}{2} \right\|^2 \leqslant \tfrac{1}{2} \|g_n\|^2 + \tfrac{1}{2} \|g_m\|^2 - \delta^2.$$

Letting m and n increase without bound, we observe that the right side of the last inequality approaches zero; since the left side assumes only non-negative values, it must also approach zero. Hence the sequence $g_1, g_2, g_3, \ldots$ is Cauchy; since the space H is complete, this sequence must converge to a vector g in H, and since S is closed, $g \in S$. We know that, given any convergent sequence in a normed linear space, the limit of the norms exists and equals the norm of this limit. Therefore, $\|g\| = \delta$; thus, there exists a vector g in S such that $\|g\| \leqslant \|h\|$, where h is any vector in S. However, if h is any vector in S other than g, $\|h\|$ must exceed δ, for if $\|h\|$ were also equal to δ the parallelogram law would furnish the equality

$$\left\| \frac{g + h}{2} \right\|^2 = \tfrac{1}{2} \|g\|^2 + \tfrac{1}{2} \|h\|^2 - \left\| \frac{g - h}{2} \right\|^2 = \delta^2 - \left\| \frac{g - h}{2} \right\|^2$$

and hence $\|\frac{1}{2}(g + h)\| < \delta$. Taking account of the convexity of S, we see that this would imply the existence of a member of S whose distance from o is *strictly less* than δ, contradicting the definition of δ. Thus, the proof is complete.

Remarks: (*a*) The reader may find it helpful to make a drawing which illustrates (in the plane) the geometrical interpretation of the preceding arguments.

(*b*) The full force of the convexity assumption has not been used. We have only employed the particular case that $\alpha = 1 - \alpha = \frac{1}{2}$. However, the assumption that S is closed cannot be omitted. (Cf. Exercise 1.)

(*c*) It is reasonable to ask whether Theorem 1 continues to hold in an arbitrary Banach space. That the answer is negative is easily seen by reference to Figure 2 and the accompanying discussion. If the l^∞ — norm is employed, it is evident that the distance from the vector $(2, 0)$ to the closed unit ball is equal to unity, and that all points on the right edge of the ball (i.e., all vectors of the form $(1, \alpha)$, $-1 \leqslant \alpha \leqslant 1$) are at unit distance from the given vector. Thus, the *uniqueness* part of Theorem 1 certainly fails in Banach spaces, while a more delicate example (in an infinite-dimensional Banach space) shows that the *existence* part of the theorem also fails. The reader should find the search for such an example highly instructive.

Corollary (Projection Theorem): *Let S be a subspace of a Hilbert space H. Then every vector f in H can be expressed, in a unique manner, in the form $f = g + h$, where $g \in S$ and $h \in S^\perp$.*

PROOF: S is non-empty, closed, and convex, and so the preceding theorem guarantees the existence of a unique vector g in S such that $\|f - g\| < \|f - \tilde{g}\|$ for every vector $\tilde{g}$ in S other than g. Let g_1 be any vector in S and let ϵ be any real number. Then $g + \epsilon g_1 \in S$, and so

$$\|f - g\| \leqslant \|f - (g + \epsilon g_1)\|, \quad \text{or} \quad \|f - g\|^2 \leqslant \|(f - g) - \epsilon g_1\|^2.$$

Expanding out the right side (as an inner product) and recalling the restriction of ϵ to real values, we obtain $\|f - g\|^2 \leqslant \|f - g\|^2 - 2\epsilon \operatorname{Re}(f - g, g_1) + \epsilon^2(g_1, g_1)$, or $\epsilon\{\epsilon(g_1, g_1) - 2 \operatorname{Re}(f - g, g_1)\} \geqslant 0$. Now suppose that $\operatorname{Re}(f - g, g_1) > 0$. Then, by continuity, the quantity in { } is *negative* for ϵ positive and sufficiently small. Thus, the left side would become, for such a choice of ϵ, the product of a positive and a negative factor, contradicting the inequality. Therefore, we have eliminated the possibility that $\operatorname{Re}(f - g, g_1) > 0$; similarly, $\operatorname{Re}(f - g, g_1)$ cannot be negative, and so $\operatorname{Re}(f - g, g_1) = 0$. Since g_1 may be replaced by ig_1, we obtain $\operatorname{Re}(f - g, ig_1) = 0$, or $\operatorname{Im}(f - g, g_1) = 0$, and hence $(f - g, g_1) = 0$. (If H is a space over the real field, the latter argument is simply omitted.) Thus, $f - g$ is orthogonal to every vector belonging to S; setting $f - g$ equal to h, we have the desired representation of f.

We conclude this section by discussing in some detail a simple, but not utterly trivial, pɪoblem closely related to Theorem 1. In the following chapter we shall discuss a number of significant extensions of this problem.

Let H be the finite-dimensional complex inner-product space consisting of all polynomials of degree $\leqslant 100$, restricted to the interval $[-1, 1]$, with inner product $(f, g) = \int_{-1}^{1} f\bar{g}$. (Needless to say, the interval $[-1, 1]$ and the number 100 can be replaced by any bounded interval and any positive integer, respectively.) We know by Theorem 2–5 that H is complete, for the functions $1, x, x^2, \ldots, x^{100}$ constitute a basis. Let a be any specified point of $[-1, 1]$ and let S be the set of all members of H satisfying $f(a) = 1$. Clearly, S is non-empty (it contains the function $f \equiv 1$) and convex (for if $f(a) = g(a) = 1$, then $\alpha f(a) + (1 - \alpha)g(a) = 1$, even without the restriction that α be real and $0 \leqslant \alpha \leqslant 1$).

We now proceed to establish the remaining hypothesis of Theorem 1, namely, that S is closed. Suppose, therefore, that $f_1, f_2, f_3, \ldots$ is any Cauchy sequence in S. Since H is complete we know that this sequence must converge to some vector f—i.e., there exists a vector f such that $\|f - f_n\| \to 0$. It remains to show that f belongs to S. Consider the collection of functions $\{(x - a), (x - a)^2, \ldots, (x - a)^{100}, 1\}$; they obviously constitute a basis of H, and if we carry out the Gram-Schmidt process on these functions (in the indicated order) we obtain an orthonormal basis $\{g_1, g_2, \ldots, g_{100}, g_{101}\}$. Since all functions in the original basis, except the last, vanish at a, it is clear, from the nature of the Gram-Schmidt process, that $g_1(a) = g_2(a) = \cdots = g_{100}(a) = 0$. Furthermore, $g_{101}(a) \neq 0$, for otherwise *every* linear combination of the g_k's, and hence *every* member of H, would vanish at a, which is not true (as we have already seen). Any function g in H must possess an expansion

$$g(x) = \left(\sum_{k=1}^{100} \alpha_k g_k(x) \right) + \alpha_{101} g_{101}(x),$$

and when we set $x = a$ we obtain $g(a) = \alpha_{101} g_{101}(a)$, or $\alpha_{101} = g(a)/g_{101}(a)$. Thus, $\|g\|^2 = \sum_{k=1}^{100} |\alpha_k|^2 + |g(a)/g_{101}(a)|^2 \geqslant |g(a)/g_{101}(a)|^2$. Replacing g by $f - f_n$, we obtain

$$\|f - f_n\|^2 \geqslant \frac{|f(a) - f_n(a)|^2}{|g_{101}(a)|^2} = \frac{|f(a) - 1|^2}{|g_{101}(a)|^2}.$$

Thus, $|f(a) - 1| \leqslant |g_{101}(a)| \cdot \|f - f_n\|$; letting $n \to \infty$, we see that $f(a) = 1$, and so $f \in S$. Thus, S is indeed closed.

We are now assured that S contains a uniquely determined member of smallest norm; that is, we have proved the following result: There exists a polynomial of degree $\leqslant 100$, assuming the value unity at $x = a$, which possesses smaller norm than any other such function. (Cf. Exercise 2.) In fact, from the preceding arguments we see that this *extremal function* is given by

$$f(x) = \frac{g_{101}(x)}{g_{101}(a)}$$

and that its norm is given by $\|f\|^2 = 1/|g_{101}(a)|^2$, or $\|f\| = 1/|g_{101}(a)|$.

We can, in fact, give a much more concise solution of this problem as follows. Let $\{h_1, h_2, \ldots, h_{101}\}$ be *any* orthonormal basis of H. Then every member f of H can be expressed in the form $f = \sum_{k=1}^{101} \alpha_k h_k$, $\|f\|^2$ is given by $\sum_{k=1}^{101} |\alpha_k|^2$, and the condition $f(a) = 1$ assumes the form $\sum_{k=1}^{101} \alpha_k h_k(a) = 1$. Thus, the problem of minimizing $\|f\|$, subject to the restrictions $f(a) = 1$ and $f \in H$, is equivalent to the purely algebraic problem of minimizing $\sum_{k=1}^{101} |\alpha_k|^2$ subject to the constraint $\sum_{k=1}^{101} \alpha_k h_k(a) = 1$. Referring to Example (*a*) following Definition 3–1 (with $n = 101$) and to the Schwarz inequality, we define the vectors $b = (\alpha_1, \alpha_2, \ldots, \alpha_{101})$ and $c = (\overline{h_1(a)}, \overline{h_2(a)}, \ldots, \overline{h_{101}(a)})$, and we obtain $1 = (b, c) = |(b, c)| \leqslant \|b\| \cdot \|c\|$, and hence $\|b\| \geqslant 1/\|c\|$, equality holding iff $b = \lambda c$ for some scalar λ. We determine λ by observing that $1 = (b, c) = (\lambda c, c) = \lambda(c, c)$, or $\lambda = 1/(c, c)$. Hence, the optimum choice of b is given by $b = c/(c, c)$, and this choice of b (and no other choice) furnishes the result $\|b\| = 1/\|c\|$. Returning to the space H, we see that the optimum function (i.e., the polynomial of degree $\leqslant 100$ satisfying the condition $f(a) = 1$ and possessing the smallest possible norm) is given by

$$f(x) = \frac{\sum_{k=1}^{101} \overline{h_k(a)} h_k(x)}{\sum_{k=1}^{101} |h_k(a)|^2}.$$

Furthermore, the minimum norm is given by

$$\|f\| = \left\{ \sum_{k=1}^{101} |h_k(a)|^2 \right\}^{-1/2}.$$

We emphasize that the final pair of formulas is valid for every choice of the orthonormal basis $h_1, h_2, \ldots, h_{101}$. Since f is *uniquely* determined, we obtain the remarkable result that for any two orthonormal bases of H, say $h_1, h_2, \ldots, h_{101}$ and $g_1, g_2, \ldots, g_{101}$, the identity

$$\sum_{k=1}^{101} \overline{h_k(a)} h_k(x) \equiv \sum_{k=1}^{101} \overline{g_k(a)} g_k(x)$$

must hold. Each side of this identity therefore defines a function $K(x, a)$ which is unambiguously determined by the Hilbert space H which has been under discussion. Finally, we remark that, for every function h belonging to H and for every point a in the interval $[-1, 1]$, the equality

$$h(a) = \int_{-1}^{1} h(x)\overline{K(x, a)}\, dx = \int_{-1}^{1} h(x)K(a, x)\, dx$$

must hold. (Cf. Exercise 3.)

Exercises

1. (*a*) Show that in Theorem 1 neither the assumption of convexity nor the assumption of closedness can be omitted.
   (*b*) A subset S of a linear space is said to be *mid-point convex* if for every pair of vectors $\{f, g\}$ belonging to S the vector $\frac{1}{2}(f + g)$ also belongs to S. Give a simple example of a linear space and a subset which is mid-point convex but not convex.
   (*c*) Show that a closed subset of a normed linear space which is mid-point convex is, in fact, convex.

2. Let H be the inner-product space consisting of all polynomials, restricted to the interval $[-1, 1]$, with the usual inner product. Show that H is *not* complete, and that the problem of minimizing $\|f\|$ subject to the restriction $f(a) = 1$ does not have a solution.

3. Prove the last equation of this section.

4. (*a*) Carry out the Gram-Schmidt procedure, on the interval $[-1, 1]$, on the set of functions $\{1, x, x^2, x^3\}$ (in the indicated order).
   (*b*) Carry out the same procedure on the same set of functions, but in reverse order.

§5. ORTHONORMAL BASES IN HILBERT SPACES

In this section we consider a fixed Hilbert space H, which is assumed to be separable and infinite-dimensional. (The corresponding theory for a finite-dimensional space will follow quite trivially from our discussion, while the extension of the theory to non-separable spaces requires the use of logical niceties with which we do not wish to become involved.)

Our first objective is to show that there exists a countably infinite orthonormal collection of vectors $\{h_1, h_2, h_3, \ldots\}$ such that every vector in H can be approximated arbitrarily closely by a finite linear combination of these vectors. By hypothesis, we can select a countably infinite dense subset of vectors, and we can arrange these vectors into a sequence $g_1, g_2, g_3, \ldots$; for convenience, we may assume that $g_1 \neq o$ (for otherwise we may simply interchange g_1 and g_2). We then form a subsequence of the g's in the following manner: We begin with g_1, and then we retain or remove g_2, according as the pair $\{g_1, g_2\}$ is linearly independent or linearly dependent. Then we proceed inductively, retaining g_n if it is not expressible as a linear combination of the g's which have previously been retained and rejecting g_n otherwise. The vectors which are retained are denoted $\tilde{g}_1, \tilde{g}_2, \tilde{g}_3, \ldots$. We shall see later that there are infinitely many $\tilde{g}$'s, but for the present we proceed as though this fact is not known. It is evident, from the very manner in which the $\tilde{g}$'s are selected, that they constitute a linearly independent collection of vectors, and that a vector in H is expressible as a (finite) linear combination of g's iff it is expressible as a linear combination of $\tilde{g}$'s. We now perform the Gram-Schmidt procedure on the $\tilde{g}$'s, obtaining an orthonormal collection $\{h_1, h_2, h_3, \ldots\}$. Again, by the nature of the Gram-Schmidt procedure, it is clear that every linear combination of the $\tilde{g}$'s is a linear combination of the h's, and conversely.

Now we proceed to show that there exist infinitely many h's. Suppose that there were only a finite number of h's, say $h_1, h_2, \ldots, h_n$. Given any vector f in H and any positive number ϵ, there exists some vector g_k such that $\|f - g_k\| < \epsilon$, since the g's form a dense subset of H. Since each g is expressible as a linear combination of h's, it is certainly true that g_k can be expressed in the form $\sum_{i=1}^n (g_k, h_i)h_i$ (by Theorem 3–7). Hence, $\|f - \sum_{i=1}^n \alpha_i h_i\| < \epsilon$ for some choice of the α's, and it now follows from Theorem 3–9 that $\|f - \sum_{i=1}^n (f, h_i)h_i\| < \epsilon$. Since n is fixed and ϵ is arbitrarily small, it follows that $f = \sum_{i=1}^n (f, h_i)h_i$. Since f is an arbitrary vector in H, it then follows that H is n-dimensional, contradicting our assumption that H is infinite-dimensional. Thus, the vectors $h_1, h_2, h_3, \ldots$ constitute a countably infinite orthonormal collection.

Now, by choosing a sequence of positive numbers $\epsilon_1, \epsilon_2, \epsilon_3, \ldots$ converging to zero, we conclude, by an obvious extension of the reasoning employed in the previous paragraph, that $\lim_{n\to\infty} \|f - \sum_{i=1}^n (f, h_i)h_i\| = 0$, or $\lim_{n\to\infty} \sum_{i=1}^n (f, h_i)h_i = f$. From this, in turn, we obtain the equality $\|f\|^2 = \sum_{i=1}^\infty |(f, h_i)|^2$, which is known as *Parseval's relation* and should be compared carefully with Bessel's inequality.

While it is *not* true that every vector in H can be expressed as a *finite* linear combination of the h's, we nevertheless refer to the collection of h's as a basis (more accurately, an orthonormal basis) of H; we thus contradict the definition of a basis given in §1, but this minor offense causes no difficulty.

If in the equality $\|f\|^2 = \sum_{i=1}^\infty |(f, h_i)|^2$ we replace f in turn by $f + g$, $f - g, f + ig, f - ig$ and employ part (*b*) of Theorem 3–3, we obtain the following equality, which is also known as *Parseval's relation:* $(f, g) = \sum_{i=1}^\infty (f, h_i)\overline{(g, h_i)}$. (Of course, the second form of this relation reduces to the first when g is replaced by f.) It should be evident (we dispense with a formal proof) that an orthonormal collection of vectors $\{h_1, h_2, \ldots\}$ constitutes a basis of H iff the Parseval relation holds for every vector (or pair of vectors) in H, and that any orthonormal collection of vectors which fail to constitute a basis can be enlarged to a basis. Also, an orthonormal collection of vectors constitutes a basis iff o is the only vector orthogonal to each member of the collection.

It is interesting to note that the completeness of H has not been used up to this point. However, the hypothesis of completeness is needed in order to prove the following important result, which may be considered as a converse of the Parseval relation.

Theorem I: *Let the vectors $h_1, h_2, h_3, \ldots$ constitute an orthonormal basis in H and let the scalars $\alpha_1, \alpha_2, \alpha_3, \ldots$ be given. Then there exists a vector f such that $(f, h_i) = \alpha_i$ for all indices i iff the series $\sum_{i=1}^{\infty} |\alpha_i|^2$ converges (to a finite sum); when this condition is satisfied the vector f is uniquely determined and is given by the relation*

$$f = \sum_{i=1}^{\infty} \alpha_i h_i = \lim_{n \to \infty} \sum_{i=1}^{n} \alpha_i h_i .$$

PROOF: If such a vector f exists, the convergence of the series $\sum_{i=1}^{\infty} |\alpha_i|^2$ $(= \sum_{i=1}^{\infty} |(f, h_i)|^2)$ is assured by Bessel's inequality. Conversely, if the series $\sum_{i=1}^{\infty} |\alpha_i|^2$ converges, let the vectors f_n be defined in the obvious manner: $f_n = \sum_{i=1}^{n} \alpha_i h_i$, $1 \leqslant n < \infty$. Then whenever $n > m$ the equality $\|f_n - f_m\|^2 = \sum_{i=m+1}^{n} |\alpha_i|^2$ holds, and the assumed convergence of the series $\sum_{i=1}^{\infty} |\alpha_i|^2$ guarantees that the sequence $f_1, f_2, f_3, \ldots$ is Cauchy. Since H is complete, there exists a vector f such that $\|f - f_n\| \to 0$. Given any index k, we obtain for any index n greater than k the equality $(f, h_k) = (f_n, h_k) + (f - f_n, h_k) = \sum_{i=1}^{n} \alpha_i (h_i, h_k) + (f - f_n, h_k) = \alpha_k + (f - f_n, h_k)$, or $(f, h_k) - \alpha_k = (f - f_n, h_k)$. Employing the Schwarz inequality and letting n increase without bound, we obtain $|(f, h_k) - \alpha_k| \leqslant \|f - f_n\| \cdot \|h_k\| = \|f - f_n\| \to 0$, and so $(f, h_k) = \alpha_k$. Finally, the uniqueness of f is easily established as follows. If g is any vector satisfying the prescribed conditions, then $(f - g, h_k) = 0$ for all indices k, and from the Parseval relation we obtain $\|f - g\|^2 = 0$, or $f = g$.

We now observe that this theorem may be restated in the following manner: There exists a one-to-one correspondence between H and l^2, the vector f in H corresponding to the vector $(\alpha_1, \alpha_2, \alpha_3, \ldots)$ in l^2, where $\alpha_i = (f, h_i)$. Furthermore, this correspondence obviously preserves the vector-space operations in H and l^2 and also preserves inner products, for if the vectors f and g in H correspond to the vectors $(\alpha_1, \alpha_2, \alpha_3, \ldots)$ and $(\beta_1, \beta_2, \beta_3, \ldots)$ in l^2, respectively, then $(f, g) = \sum_{i=1}^{\infty} \alpha_i \bar{\beta}_i$. Thus, we have established an isomorphism between H and l^2, and from this it follows readily that any two separable infinite-dimensional Hilbert spaces are isomorphic, so that, in a sense, there is only one such Hilbert space. It should be emphasized, however, that the correspondence between two separable infinite-dimensional Hilbert spaces is not uniquely determined, for there is wide freedom in the selection of an orthonormal basis in each space.

The preceding result, despite its elegance, may be misleading, for in many practical problems one may find it necessary to work with a specific Hilbert space, and, while the existence of an unlimited number of orthonormal bases is assured, one may wish to find a particularly useful basis. We discuss this problem briefly for three particular spaces.

(*a*) In l^2 the most obvious orthonormal basis, and the most convenient for use in many problems, is furnished by the vectors $(1, 0, 0, 0, \ldots)$, $(0, 1, 0, 0, \ldots)$, $(0, 0, 1, 0, \ldots)$, $\ldots$.

(*b*) Let A denote the interval $[-1, 1]$, and let $h_1, h_2, h_3, \ldots$ be the functions obtained by performing the Gram-Schmidt process on the functions $1, x, x^2, x^3, \ldots$. (Cf. Exercise 4–4.) In this manner we evidently obtain a sequence of polynomials, the polynomial h_n being of degree $n - 1$. Now, from Theorem 7–2 of Chapter 3 we know that, given any function f belonging to $L^2(A)$ and any positive number ϵ, there exists a polynomial p such that $\|f - p\|_2 < \epsilon$. From the fact that the collection $\{h_1, h_2, h_3, \ldots\}$ contains exactly one polynomial of each degree (including degree zero) it is evident that p can be expressed as a linear combination of the h's; in fact, if p is of degree k it is clear that p is expressible as a combination of the first $k + 1$ h's. Hence, the linear combinations of the h's form a dense subset of $L^2(A)$, and, therefore, the h's form an orthonormal basis of $L^2(A)$.

(*c*) The functions $1, e^{2\pi ix}, e^{-2\pi ix}, e^{4\pi ix}, e^{-4\pi ix}, \ldots$ form an orthonormal collection on the interval $[0, 1]$. Again referring to Theorem 7–2 of Chapter 3, we can, given any function f belonging to $L^2([0, 1])$ and any positive number ϵ, find a continuous function g on this interval such that $\|f - g\|_2 < \epsilon/3$; by modifying g appropriately at one end of the interval, if necessary, we can obtain a continuous function $\tilde{g}$ such that $\tilde{g}(0) = \tilde{g}(1)$ and $\|g - \tilde{g}\|_2 < \epsilon/3$; finally, as shown in Appendix D, there exists a (finite) linear combination, which we denote as t, of the given functions, such that $|\tilde{g}(x) - t(x)| < \epsilon/3$ on $[0, 1]$. (The symbol t is employed to suggest "trigonometric," since functions of this form are known as trigonometric polynomials.) Hence, $\|\tilde{g} - t\|_2^2 = \int_0^1 |\tilde{g} - t|^2 < (\epsilon/3)^2$, or $\|\tilde{g} - t\|_2 < \epsilon/3$. By the triangle inequality we obtain $\|f - t\|_2 < \epsilon$, and from this result we conclude, exactly as in (*b*), that the given collection of functions form an orthonormal basis in $L^2([0, 1])$. Since this collection of functions was the first orthonormal collection to be thoroughly studied, beginning with Fourier, the quantities (f, h_k) appearing in the representation of a vector f in any Hilbert space as a series in the orthonormal vectors $h_1, h_2, h_3, \ldots$ are frequently called *Fourier coefficients*.

It should be emphasized, in connection with (*b*) and (*c*), that there is no assurance that the series expansions of a given function in terms of a particular orthonormal collection converge pointwise; only convergence in the L^2-norm is guaranteed. We shall return briefly to this topic in §5–2.

Exercises

1. Prove the projection theorem (in a separable Hilbert space) by introducing an orthonormal basis in the subspace upon which the projection is performed.

2. Let $p_1(x) = 1$, $p_2(x) = (d/dx)(x^2 - 1)$, $p_3(x) = d^2(x^2 - 1)^2/dx^2$, ... Prove that $\int_{-1}^{1} p_i p_j \,(= \int_{-1}^{1} p_i \bar{p}_j = (p_i, p_j))$ equals zero if $i \neq j$. (From this it follows easily that these polynomials are constant multiples of the polynomials described in the discussion of $L^2([-1, 1])$.) Show that $p_i(1) \neq 0$ for each index i, so that the polynomials $p_i(x)/p_i(1)$ also constitute an orthogonal collection over the interval $[-1, 1]$. (These are the Legendre polynomials; it is standard practice to index them according to their degree, so that $P_0(x) = p_1(x)/p_1(1)$, $P_1(x) = p_2(x)/p_2(1), \ldots$.)

3. Let $f(x) = x - \frac{1}{2}$ in the interval $[0, 1]$. Determine the Fourier coefficients of this function with respect to the orthonormal basis discussed in (c), and then, by invoking the Parseval relation, evaluate the sum of the series $\sum_{k=1}^{\infty} 1/k^2$.

CHAPTER 5

LINEAR FUNCTIONALS

The major part of the theory of linear spaces consists of the study of mappings of one such space into another, particularly linear mappings. In this chapter we shall be concerned with mappings of a given linear space into its field of scalars; such mappings are called *functionals.* After presenting the basic ideas, we shall develop to some extent the theory of linear functionals when the linear space is normed. As in previous chapters, we deal with real and complex spaces impartially, except at a few places where there is in fact some essential difference between the two cases.

§1. BASIC DEFINITIONS AND CONCEPTS

Definition 1: *Let L be any linear space, and let the scalar $l(f)$ be defined for each vector f in L. Then the function l (whose domain is L and whose range is part or all of the field of scalars) is termed a functional (on L). If for every pair of vectors $\{f, g\}$ and for every scalar α the equalities $l(f + g) = l(f) + l(g)$ and $l(\alpha f) = \alpha l(f)$ hold, then l is termed a linear functional.*

Examples

(*a*) Let L be the class of all (real- or complex-valued) functions defined on the interval $[0, 1]$, and let $l(f) = f^2(\frac{1}{2})$. Then l is a functional on L, but it is not linear.

(*b*) Let L be as in (*a*), with $l(f) = 2f(\frac{1}{2}) - 3f(\frac{2}{3})$. Then l is a linear functional on L.

Theorem 1: *(a) If l is any linear functional, $l(o) = 0$.*

(b) If l is any linear functional, if $f_1, f_2, f_3, \ldots, f_n$ are any vectors, and if $\alpha_1, \alpha_2, \ldots, \alpha_n$ are any scalars, then $l(\alpha_1 f_1 + \alpha_2 f_2 + \cdots + \alpha_n f_n) = \alpha_1 l(f_1) + \alpha_2 l(f_2) + \cdots + \alpha_n l(f_n)$.

PROOF: Trivial.

Theorem 2: *Let l be a linear functional defined on a finite-dimensional linear space, let the vectors $f_1, f_2, \ldots, f_n$ constitute a basis, and let the scalars $l(f_1), l(f_2), \ldots, l(f_n)$ be known. Then l is completely determined. Conversely, if scalars $\alpha_1, \alpha_2, \ldots, \alpha_n$ are chosen at pleasure, there exists one and only one linear functional l such that $l(f_k) = \alpha_k$, $1 \leqslant k \leqslant n$.*

PROOF: Every vector f possesses a *unique* representation of the form $f = \beta_1 f_1 + \beta_2 f_2 + \cdots + \beta_n f_n$, and so $l(f)$ must equal $\beta_1 l(f_1) + \beta_2 l(f_2) + \cdots + \beta_n l(f_n)$. Conversely, for every vector f let $l(f) = \sum_{k=1}^{n} \alpha_k \beta_k$. Then it is trivially evident that l is a linear functional satisfying $l(f_k) = \alpha_k$, $1 \leqslant k \leqslant n$.

Definition 2: *Let l be any functional (not necessarily linear) on L and let α be any scalar. Then αl is the functional defined in the obvious manner: $(\alpha l)(f) = \alpha \cdot l(f)$. If l_1 and l_2 are any two functionals on L, the functional $l_1 + l_2$ is also defined in the obvious manner: $(l_1 + l_2)(f) = l_1(f) + l_2(f)$.*

Theorem 3: *The class of all linear functionals defined on a given linear space L constitute a linear space (with the definitions of addition and multiplication by scalars given in Definition 2).*

PROOF: Trivial.

Definition 3: *Let N be a normed linear space and let l be a linear functional on N. Then l is said to be bounded if there exists a real number C (necessarily $\geqslant 0$) such that $|l(f)| \leqslant C \|f\|$ for every vector $f \in N$. If l is a bounded linear functional, its norm is defined as the greatest lower bound of all numbers C for which the preceding inequality holds; the norm of a bounded linear functional, l, will be denoted $\|l\|$.*

Remark: The use of the symbol $\| \;\; \|$ for the norm of a bounded linear functional (and the very use of the word "norm") suggests very strongly, in conjunction with Theorem 3, that the bounded linear functionals on N constitute a normed linear space closely related to N; it will be seen later, in the corollary to Theorem 4, that this is indeed the case.

Examples

(*a*) Let N be the class of all (real-valued or complex-valued) continuous functions defined on the compact interval $[a, b]$, and let $\|f\| = \max_{a \leqslant x \leqslant b} |f(x)|$. If $l(f) = f(c)$, where c is any fixed point on the interval $[a, b]$, then l is a bounded linear functional, and $\|l\| = 1$.

(*b*) Let N be the same class of functions as in (*a*), let $l(f)$ be defined as in (*a*), but let $\|f\| = \int_a^b |f(x)|\, dx$. Then l is, of course, still a linear functional, but it is not bounded, for the ratio $|f(c)|/\|f\|$ can evidently be made arbitrarily large.

(*c*) Let N be any inner-product space, let g be any fixed vector in N, and let $l(f) = (f, g)$ for all f in N. Then l is a bounded linear functional, and $\|l\| = \|g\|$. (Cf. Theorem 3–1.)

Theorem 4: *If l_1 and l_2 are bounded linear functionals defined on the normed linear space N and if α is any scalar, then $\|\alpha l_1\| = |\alpha| \cdot \|l_1\|$ and $\|l_1 + l_2\| \leqslant \|l_1\| + \|l_2\|$.*

PROOF: Trivial.

Corollary: *Let N be a normed linear space and let N^*, the dual space of N, be the space consisting of all bounded linear functionals on N, normed in accordance with Definition* 3. *Then N^* is complete, even if N is not complete.*

PROOF: Left to reader as Exercise 1.

Theorem 5: *A linear functional l defined on a normed linear space is continuous iff l is bounded, and also iff l is continuous at o.*

PROOF: Left to reader as Exercise 3.

Exercises

1. Prove the corollary of Theorem 4.
2. Prove that a linear functional defined on a finite-dimensional normed linear space is automatically bounded.
3. Prove Theorem 5.

§2. THE PRINCIPLE OF UNIFORM BOUNDEDNESS

This section is devoted to a single result, the Banach-Steinhaus theorem, or principle of uniform boundedness, together with an application of this result to an interesting problem of classical analysis.

Theorem 1: *Let B be a Banach space and let $\{l_\alpha\}$ be a (non-empty) collection of bounded linear functionals on B. Suppose that the l_α's are pointwise bounded—i.e., for each $f \in B$ there exists a number K such that $|l_\alpha(f)| \leqslant K$ for all indices α. (Of course, K will vary, in general, with f; the essential point is that K does not depend on α.) Then the l_α's are bounded in norm—i.e., there exists a constant C such that $\|l_\alpha\| \leqslant C$ for all indices α.*

PROOF: For each positive integer n let $E_{n,\alpha}$ be the subset of B on which $|l_\alpha(f)| \leqslant n$. Since l_α is continuous (by Theorem 1–5), $E_{n,\alpha}$ is closed. Hence the set $A_n = \bigcap_\alpha E_{n,\alpha}$ is also closed. (Note that A_n is the set of all vectors f for which all the numbers $|l_\alpha(f)|$ are $\leqslant n$.) By hypothesis, every f belongs to at least one A_n—e.g., if $|l_\alpha(f)| \leqslant 2.7$ for all indices α, then $f \in A_3$. Thus, $\bigcup_n A_n = B$, and since B is complete, the Baire category theorem guarantees that at least one of the closed sets A_n, say A_N, has non-empty interior; i.e., there exists a vector x_0 and a positive number r such that, for every index α, $|l_\alpha(x)| \leqslant N$ whenever $\|x - x_0\| \leqslant r$. Letting $y = x - x_0$, we obtain $|l_\alpha(y + x_0)| \leqslant N$ whenever $\|y\| \leqslant r$. Since $l_\alpha(y) = l_\alpha(y + x_0) - l_\alpha(x_0)$, we obtain $|l_\alpha(y)| \leqslant |l_\alpha(y + x_0)| + |l_\alpha(x_0)| \leqslant N + N = 2N$ whenever $\|y\| \leqslant r$. For any f except o we may write $f = (\|f\|/r)(r/\|f\|)f$, and so

$$|l_\alpha(f)| = \frac{\|f\|}{r}\left|l_\alpha\left(\frac{r}{\|f\|}f\right)\right| \leqslant \frac{\|f\|}{r} \cdot 2N,$$

since $\|(r/\|f\|)f\| = r$. Thus, for any vector f (the case $f = o$ is trivial) and for any index α, we obtain $|l_\alpha(f)| \leqslant (2N/r)\,\|f\|$, and so $\|l_\alpha\| \leqslant 2N/r$ for all α.

We now proceed to apply the preceding theorem to the theory of Fourier series.* We shall state here only those elementary ideas which will be employed in the following development. Let $P[-\pi, \pi]$ denote the class of continuous functions (either real-valued or complex-valued) defined on the closed interval $[-\pi, \pi]$ satisfying $f(-\pi) = f(\pi)$ (so that f can be extended continuously to all of R as a function with period 2π). Clearly $P[-\pi, \pi]$ becomes a Banach space if $\|f\|$ is defined as max $|f(x)|$ for every $f \in P[-\pi, \pi]$. With each member f of the collection $P[-\pi, \pi]$ we associate the Fourier series

$$\tfrac{1}{2}a_0 + \sum_{k=1}^{\infty}(a_k \cos kx + b_k \sin kx),$$

---

* In contrast to the discussion in §4–5, we find it slightly more convenient here to work with the interval $[-\pi, \pi]$ rather than $[0, 1]$ and to employ the trigonometric functions rather than the exponential functions.

where

$$a_k = \frac{1}{\pi}\int_{-\pi}^{\pi} f(t)\cos kt\,dt \quad \text{and} \quad b_k = \frac{1}{\pi}\int_{-\pi}^{\pi} f(t)\sin kt\,dt.$$

The most obvious questions concerning this series are: (1) Does it converge and (2) if it does converge, is the sum of the series equal to $f(x)$? (Of course, it is conceivable that the series converges for some values of x and not for some other values of x.) The answer to the second question is affirmative—*if* the series converges for a particular value of x it converges correctly—that is, it converges to $f(x)$. The answer to the first question is much more complicated. For example, it is easy to prove that if f is *differentiable* at a particular value of x, then the series converges for that value of x. With more effort, it can be shown that the series converges at x if the function f is *monotone* in some neighborhood of x, however small. As progress was made in proving convergence with weaker and weaker hypotheses, it was wondered whether, in fact, the series must converge without *any* hypotheses on f (except the initial hypothesis of continuity). The answer to this is *no*—we shall proceed to prove that there exists a member of $P[-\pi, \pi]$ whose Fourier series *diverges* at $x = 0$. (The choice $x = 0$ is purely a matter of convenience—any other value of x may be handled in a similar manner.)

When $x = 0$ the Fourier series simplifies to $\frac{1}{2}a_0 + a_1 + a_2 + \cdots$. The zeroth partial sum is $\frac{1}{2}a_0$, the first partial sum is $\frac{1}{2}a_0 + a_1$, etc. Denoting these partial sums by $l_0, l_1, l_2, \ldots$ we obviously obtain (by referring to the preceding definition of the a_n's)

$$l_k(f) = \frac{1}{\pi}\int_{-\pi}^{\pi} f(t)\{\tfrac{1}{2} + \cos t + \cos 2t + \cdots + \cos kt\}\,dt.$$

Obviously l_k is a linear functional; also, it is bounded, since

$$|l_k(f)| \leqslant \frac{1}{\pi}\int_{-\pi}^{\pi} (\max|f(t)|)\cdot(\tfrac{1}{2} + 1 + 1 + \cdots + 1) = (2k + 1)\,\|f\|.$$

Thus, $\|l_k\|$ is certainly not more than $(2k + 1)$. However, we shall need a more accurate estimate on $\|l_k\|$. To obtain this estimate, we employ the identity (cf. Exercise 1)

$$\tfrac{1}{2} + \cos t + \cos 2t + \cdots + \cos kt = \frac{\sin(k + \frac{1}{2})t}{2\sin\frac{1}{2}t}, \qquad k = 1, 2, 3, \ldots.$$

Thus,

$$|l_k(f)| \leqslant \int_{-\pi}^{\pi} |f(t)|\cdot\left|\frac{\sin(k + \frac{1}{2})t}{2\pi\sin\frac{1}{2}t}\right| dt \leqslant \|f\|\cdot\int_{-\pi}^{\pi}\left|\frac{\sin(k + \frac{1}{2})t}{2\pi\sin\frac{1}{2}t}\right| dt,$$

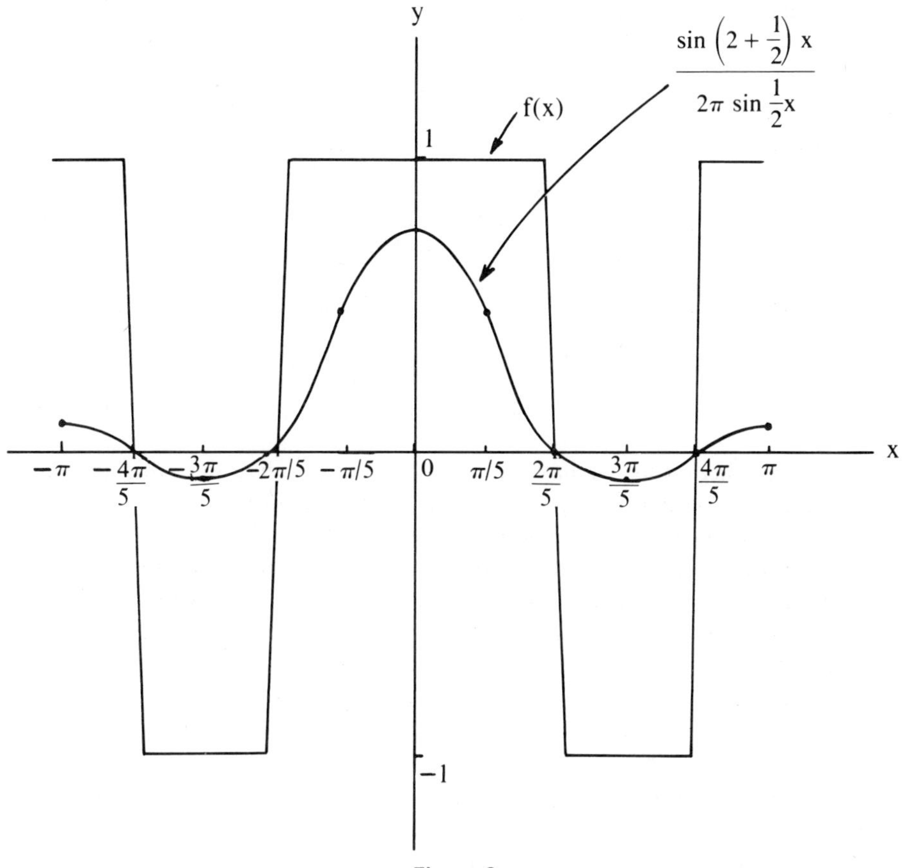

Figure 3

and so

$$\|l_k\| \leqslant \int_{-\pi}^{\pi} \left| \frac{\sin (k + \frac{1}{2})t}{2\pi \sin \frac{1}{2}t} \right| dt.$$

We shall now demonstrate that, in fact, equality must hold; we do this by demonstrating a function f (belonging to $P[-\pi, \pi]$) of unit norm such that

$$|l_k(f)| \geqslant -\epsilon + \int_{-\pi}^{\pi} \left| \frac{\sin (k + \frac{1}{2})t}{2\pi \sin \frac{1}{2}t} \right| dt,$$

where ϵ is any prescribed positive number. Consider, for ease of exposition, the particular case $k = 2$, and let the function f be chosen as indicated in Figure 3. Clearly, f is a member of $P[-\pi, \pi]$, and if the inclined segments are very steep, the integral

$$\int_{-\pi}^{\pi} f(t) \frac{\sin (k + \frac{1}{2})t}{2\pi \sin \frac{1}{2}t} dt$$

is very close to

$$\int_{-\pi}^{\pi} \left| \frac{\sin (k + \frac{1}{2})t}{2\pi \sin \frac{1}{2}t} \right| dt,$$

and so, for this particular f, $l_k(f)$ is very close to

$$\int_{-\pi}^{\pi} \left| \frac{\sin (k + \frac{1}{2})t}{2\pi \sin \frac{1}{2}t} \right| dt.$$

Thus, the norm of the functional l_k is *at least*

$$\int_{-\pi}^{\pi} \left| \frac{\sin (k + \frac{1}{2})t}{2\pi \sin \frac{1}{2}t} \right| dt;$$

on the other hand, we have seen that $\|l_k\|$ is *not more* than this amount. Hence, $\|l_k\|$ equals this quantity, as asserted previously.

Having obtained an exact expression for $\|l_k\|$ in the form of a definite integral, we can now easily show that $\|l_k\|$ is an unbounded function of k; in fact, $\|l_k\| \to +\infty$ as k increases without bound. To demonstrate this fact, we rewrite our previous result in the form

$$\|l_k\| = \frac{1}{\pi} \int_0^{\pi} \frac{|\sin (k + \frac{1}{2})t|}{t} \cdot \frac{t}{\sin \frac{1}{2}t} \, dt.$$

For t in $[0, \pi]$, $t/(\sin \frac{1}{2}t)$ is continuous and positive; since this interval is compact, the minimum of $t/(\sin \frac{1}{2}t)$ is a strictly positive number, say α. (Actually, $\alpha = 2$, but this is of no importance.) Therefore,

$$\|l_k\| > \frac{\alpha}{\pi} \int_0^{\pi} \frac{|\sin (k + \frac{1}{2})t|}{t} \, dt = \frac{\alpha}{\pi} \int_0^{(k+1/2)\pi} \frac{|\sin u|}{u} \, du.$$

Hence (for $k > 0$),

$$\|l_k\| > \frac{\alpha}{\pi} \int_0^{k\pi} \frac{|\sin u|}{u} \, du = \frac{\alpha}{\pi} \sum_{m=0}^{k-1} \int_{m\pi}^{(m+1)\pi} \frac{|\sin u|}{u} \, du$$

$$> \frac{\alpha}{\pi} \sum_{m=0}^{k-1} \frac{1}{(m+1)\pi} \int_{m\pi}^{(m+1)\pi} |\sin u| \, du = \frac{2\alpha}{\pi^2} \sum_{m=0}^{k-1} \frac{1}{m+1}.$$

More explicitly,

$$\|l_k\| > \frac{2\alpha}{\pi^2} \left(1 + \tfrac{1}{2} + \cdots + \frac{1}{k} \right),$$

and so $\|l_k\| \to +\infty$ as k increases without bound, since the harmonic series $1 + \frac{1}{2} + \frac{1}{3} + \cdots$ is known to diverge.

Now suppose that for *every* f in $P[-\pi, \pi]$ the Fourier series of f converges at $x = 0$. This means that the sequence of scalars $l_0(f)$, $l_1(f)$, $l_2(f)$, . . . converges. Since a convergent sequence of scalars is certainly bounded, the family of bounded linear functionals $\{l_k\}$ (where $k = 0$, 1, 2, . . .) would be *pointwise* bounded, and then, according to Theorem 1, the set of numbers $\{\|l_k\|\}$ would be bounded, contradicting the result $\|l_k\| \to +\infty$ which we have just demonstrated. Hence, there must exist a function f in $P[-\pi, \pi]$ whose Fourier series diverges at $x = 0$.

Remarks: (*a*) Note that the preceding argument does *not* tell us how to find a function f whose Fourier series diverges at $x = 0$; it only assures us that such a function exists. A specific example is presented in [Titchmarch, pp. 416–418].

(*b*) We have done *more* than show that there exists a function whose Fourier series diverges at $x = 0$; we have shown that there exists a function whose Fourier series diverges *unboundedly* at $x = 0$.

(*c*) By a slightly more refined argument we can even show that there exists a function in $P[-\pi, \pi]$ whose Fourier series diverges on some *uncountable dense* subset of $[-\pi, \pi]$. (See [Rudin, pp. 101–103].)

(*d*) Part (c) *suggests* that there exists a function in $P[-\pi, \pi]$ whose Fourier series diverges *everywhere* in $[-\pi, \pi]$. The answer to this was not found until 1966, when Carleson showed that the Fourier series of any function in $L^2([-\pi, \pi])$, and hence of any function in $P[-\pi, \pi]$, converges almost everywhere. On the other hand, it was shown much earlier, by Kolmogoroff, that there exists a function in $L^1([-\pi, \pi])$ whose Fourier series diverges everywhere. (Since the functions $\cos x$, $\sin x$, $\cos 2x$, $\sin 2x$, . . . are bounded, the coefficients $a_1, b_1, a_2, b_2, \ldots$ are defined for any function f in $L^1([-\pi, \pi])$, and so one can associate a Fourier series with f.)

Exercises

1. Prove the trigonometric identity previously stated, namely:

$$\tfrac{1}{2} + \cos t + \cos 2t + \cdots + \cos kt = \frac{\sin (k + \frac{1}{2})t}{2 \sin \frac{1}{2}t}.$$

2. Prove that $\min_{0 \leqslant t \leqslant \pi} \dfrac{t}{\sin \frac{1}{2}t} = 2.$

§3. BOUNDED LINEAR FUNCTIONALS IN HILBERT SPACES

The task of identifying in some reasonably explicit manner the set of all bounded linear functionals on a given normed linear space is in general a very difficult one; in the case of the spaces $L^p(A)$ a very satisfying and elegant solution to this problem does exist, as we shall indicate in the following section. The present section is devoted to obtaining the surprisingly simple solution to this problem in the case of a Hilbert space and to one application.

Theorem I (Riesz Representation Theorem):* *Let l be any bounded linear functional defined on a Hilbert space H. Then there exists a unique vector in H, which we shall denote g_l, such that $l(f) = (f, g_l)$ for every vector f. Furthermore, $\|l\| = \|g_l\|$.*

PROOF: First, uniqueness is easily established, for if $g \neq h$, then $(g - h, g - h) > 0$, and so $(g - h, g) \neq (g - h, h)$; this shows that g and h cannot determine the same bounded linear functional. Secondly, if l is the zero functional, the zero vector o will suffice. Setting aside this trivial case, let S denote the null-space of l—i.e., the set of all vectors f such that $l(f) = 0$. Since l is linear, S is a linear manifold, and since l is continuous, S must be closed and, hence, a subspace of H. Since l is not the zero functional, there exist vectors f such that $l(f) \neq 0$. We choose any such vector, say f_1, and express it, in accordance with the projection theorem, in the form

$$f_1 = g_1 + h_1 \ (g_1 \in S, h_1 \in S^\perp, h_1 \neq o).$$

Now let f be any vector in H. Appealing once again to the projection theorem, we write

$$f = g + h \ (g \in S, h \in S^\perp)$$

Now let us consider the vector $\tilde{f} = f - \{l(f)/l(f_1)\}f_1$. By linearity, $l(\tilde{f}) = l(f) - \{l(f)/l(f_1)\}l(f_1) = 0$, and therefore $\tilde{f} \in S$. On the other hand,

$$\tilde{f} = (g + h) - \left\{\frac{l(f)}{l(f_1)}\right\}(g_1 + h_1) = \left(g - \left\{\frac{l(f)}{l(f_1)}\right\}g_1\right) + \left(h - \left\{\frac{l(f)}{l(f_1)}\right\}h_1\right).$$

This equation clearly provides the unique decomposition of $\tilde{f}$ into the sum of a vector contained in S and a vector orthogonal to S; but since $\tilde{f} \in S$, we obtain $h - \{l(f)/l(f_1)\}h_1 = o$, or

$$l(f)h_1 = l(f_1)h.$$

* This name is also given to a much deeper theorem which provides the solution to the problem of identifying all the bounded linear functionals on the Banach space $C([a, b])$ provided with the maximum norm, $\|f\| = \max_{a \leqslant x \leqslant b} \|f(x)\|$.

Taking the inner product of both sides with h_1, we obtain

$$l(f)(h_1, h_1) = l(f_1)(h, h_1).$$

Since $(g, h_1) = 0$, we may replace the right side of this equality by $l(f_1)\{(h, h_1) + (g, h_1)\}$, $l(f_1)(f, h_1)$, or $(f, \overline{l(f_1)}h_1)$, and so

$$l(f) = \frac{1}{(h_1, h_1)}(f, \overline{l(f)}h_1) = \left(f, \frac{\overline{l(f_1)}}{(h_1, h_1)} h_1\right).$$

Letting $g_l = \overline{(l(f_1)/(h_1, h_1))}h_1$, we obtain $l(f) = (f, g_l)$. The proof that $\|l\| = \|g_l\|$ is elementary, for by the Schwarz inequality, $|l(f)| = |(f, g_l)| \leqslant \|g_l\| \cdot \|f\|$, and so $\|l\| \leqslant \|g_l\|$; on the other hand, $|l(g_l)| = (g_l, g_l) = \|g_l\| \cdot \|g_l\|$, and so $\|l\| \geqslant \|g_l\|$. Combining these two inequalities, we obtain $\|l\| = \|g_l\|$.

We shall now discuss at some length one particular application of this simple but important theorem. Let D be a bounded domain (= open non-empty connected subset) of the plane. (With trivial modifications the entire discussion can be extended to higher-dimensional euclidean spaces, and the restriction to bounded domains is made purely for ease in exposition.) A real-valued continuous function u defined on D is said to be *harmonic* in D if for every closed disc Δ contained in D the mean value of u over Δ (i.e., (area of $\Delta)^{-1} \iint_\Delta u\, dx\, dy$) equals the value of u at the center of Δ. Obviously, the class of all such functions is a *real* linear space (with the obvious definitions of addition and multiplication by scalars), and this statement remains true if we impose the additional restriction that $\iint_\Delta u^2\, dx\, dy < +\infty$. (The continuity of u and the boundedness of D fail to guarantee the finiteness of $\iint_\Delta u^2\, dx\, dy$, since u^2 may become large near the boundary of D. For example—we do not go into the details—the function $u = (1 - x)/(1 - 2x + x^2 + y^2)$ is harmonic in the open unit disc, $\{(x, y) \mid x^2 + y^2 < 1\}$, but the integral of u^2 over this domain is infinite; note that u is unbounded in the neighborhood of the boundary-point $(1, 0)$.) Furthermore, $(u, v) = \iint_\Delta uv\, dx\, dy$ constitutes a valid definition of an inner product; the inner-product space thus obtained will be denoted $H(D)$, the letter H suggesting the word "harmonic."

(The reader is probably aware that the usual definition of a harmonic function is that it is one which possesses continuous second partial derivatives in a specified domain D and satisfies *Laplace's equation*, $\partial^2 u/\partial x^2 + \partial^2 u/\partial y^2 = 0$, at all points of D. The equivalence of this definition with the one which we have presented is a result of fundamental importance in the theory of harmonic functions, but we shall have no need to refer to Laplace's equation, or even to use the fact that the functions with which we are dealing are differentiable.)

In fact, the inner-product space $H(D)$ turns out to be complete, infinite-dimensional, and separable. We now proceed to prove this assertion, although some of the details will be left as exercises. For convenience, we present the development in a sequence of lemmas.

Lemma 1: *Let $u \in H(D)$, let P be any point of D, and let R denote the distance from P to the boundary of D. (Since D is open, $R > 0$.) Then $|u(P)| \leqslant (\pi R^2)^{-1/2} \|u\|$.*

PROOF: Let Δ be a closed disc centered at P and possessing radius R', $0 < R' < R$. Then $0 \leqslant \iint_\Delta (u - u(P))^2\, dx\, dy = \iint_\Delta u^2\, dx\, dy - 2u(P) \iint_\Delta u\, dx\, dy + \iint_\Delta u^2(P)\, dx\, dy = \iint_\Delta u^2\, dx\, dy - 2u(P)\{\pi R'^2 u(P)\} + \pi R'^2 u^2(P) = \iint_\Delta u^2\, dx\, dy - \pi R'^2 u^2(P)$. Hence, $\pi R'^2 u^2(P) \leqslant \iint_\Delta u^2\, dx\, dy \leqslant \iint_D u^2\, dx\, dy = \|u\|^2$, and so $(\pi R'^2)^{1/2} |u(P)| \leqslant \|u\|$, or $|u(P)| \leqslant (\pi R'^2)^{-1/2} \|u\|$. Since R' may be chosen arbitrarily close to R, we obtain the desired inequality. (Roughly speaking, this result shows that a harmonic function possessing small norm must be pointwise small.)

Lemma 2: *If a sequence of harmonic functions $u_1, u_2, u_3, \ldots$ (not necessarily belonging to $H(D)$) converges uniformly in D, the limit function is also harmonic in D. The same result holds true if the hypothesis of uniform convergence in D is weakened to uniform convergence on every compact subset of D.*

PROOF: Left to reader as Exercise 1.

Lemma 3: *Let $u_1, u_2, u_3, \ldots$ be a Cauchy sequence in $H(D)$. Then this sequence converges pointwise in D, the limit function u is also a member of $H(D)$, and $\|u - u_n\| \to 0$ as $n \to \infty$. Thus, $H(D)$ is complete, and hence it is a Hilbert space.*

PROOF: Let A be any non-empty compact subset of D, let $r(P)$ denote the distance from any point P of D to the boundary of D, and let $R = \min_{P \in A} r(P)$. (Since A is compact, $R > 0$.) Given $\epsilon > 0$, choose N so large that whenever m and n both exceed N the inequality $\|u_n - u_m\| < (\pi R^2)^{1/2} \epsilon$ holds. Then referring to Lemma 1 (with u replaced by $u_n - u_m$) we obtain, for any point P in D, the inequality

$$|u_n(P) - u_m(P)| \leqslant (\pi r^2(P))^{-1/2} \|u_n - u_m\| < \frac{R}{r(P)} \epsilon;$$

if P is confined to A, we therefore obtain the inequality $|u_n(P) - u_m(P)| < \epsilon$. Thus, the sequence $u_1, u_2, \ldots$ converges pointwise everywhere in D and uniformly on compact subsets. By Lemma 2, the limit function, which we denote u, is harmonic in D.

Now let A again denote any compact subset of D, and let $\| \ \|^{(A)}$ be assigned a meaning analogous to that of $\| \ \|$, except that integration over A, rather than over D, is performed. Then $\|u_n\|^{(A)}$ and $\|u\|^{(A)}$ are all well-defined. (Since A is compact and u is continuous, u is bounded, and hence quadratically integrable, on A.) We then obtain $\|u\|^{(A)} \leqslant \|u - u_n\|^{(A)} + \|u_n\|^{(A)} \leqslant \|u - u_n\|^{(A)} + \|u_n\| \leqslant \|u - u_n\|^{(A)} + \sup \|u_n\|$. (Recall that the norms of the vectors constituting a Cauchy sequence are convergent, hence bounded.) Letting n increase without bound and recalling that the sequence $u_1, u_2, \ldots$ converges *uniformly* to u on A, we conclude that $\|u - u_n\|^{(A)} \to 0$, and hence $\|u\|^{(A)} \leqslant \sup \|u_n\|$. Since the right side of this inequality is independent of the particular choice of A, we can allow A to increase within D; in the limit we obtain the results that $\|u\|$ is finite and that $\|u\| \leqslant \sup \|u_n\|$.

It remains to prove that $\|u - u_n\| \to 0$ as $n \to \infty$. Let $\epsilon > 0$, let N be so chosen that $\|u_n - u_m\| < \epsilon$ whenever m and n exceed N, and let the compact subset A of D be so chosen that $\|u\|^{(D-A)} < \epsilon$ and $\|u_{N+1}\|^{(D-A)} < \epsilon$. Then, whenever $n > N$,

$$\begin{aligned}
\|u - u_n\|^2 &= \{\|u - u_n\|^{(A)}\}^2 + \{\|u - u_n\|^{(D-A)}\}^2 \\
&= \{\|u - u_n\|^{(A)}\}^2 + \{\|u - u_{N+1} + (u_{N+1} - u_n)\|^{(D-A)}\}^2 \\
&\leqslant \{\|u - u_n\|^{(A)}\}^2 + \{\|u\|^{(D-A)} + \|u_{N+1}\|^{(D-A)} + \|u_{N+1} - u_n\|\}^2 \\
&\leqslant \{\|u - u_n\|^{(A)}\}^2 + (3\epsilon)^2.
\end{aligned}$$

As before, $\|u - u_n\|^{(A)}$ approaches zero with increasing n, and so we obtain the inequality $0 \leqslant \overline{\lim} \|u - u_n\| \leqslant 3\epsilon$. Since ϵ may be chosen arbitrarily small, we obtain the desired result, namely $\|u - u_n\| \to 0$.

Lemma 4: *For each fixed point Q of D, there exists a unique member of $H(D)$, which we denote u_Q, such that for every $u \in H(D)$ the following equality holds:*

$$u(Q) = (u, u_Q) = \iint_D u u_Q \, dx\, dy.$$

PROOF: Let $l_Q(u)$ be defined, for every $u \in H(D)$, by the equation $l_Q(u) = u(Q)$; clearly, l is a linear functional, while Lemma 1 shows that l_Q is bounded. Theorem 1 now guarantees the existence and uniqueness of the function u_Q.

Lemma 5: *$H(D)$ is separable.*

PROOF: It follows immediately from Lemma 4 that a function u belonging to $H(D)$ vanishes at Q iff u is orthogonal to u_Q. Choose a countable dense subset $\{Q_1, Q_2, Q_3, \ldots\}$ of D and form the corresponding sequence of functions $u_{Q_1}, u_{Q_2}, \ldots$. A function u belonging to $H(D)$ which is orthogonal to all these functions must vanish at all the Q_k's, and then by continuity it must vanish identically. Hence, $H(D)$ is spanned by a countably infinite collection of its vectors (cf. §4–5), and is therefore separable.

Lemma 6: *$H(D)$ is infinite-dimensional.*

PROOF: Left to reader as Exercise 2.

Now, if in Lemma 4 we choose $u = u_P$, where P is any point of D, we obtain $u_P(Q) = (u_P, u_Q) = (u_Q, u_P) = u_Q(P)$. This suggests introducing the notation $K(P, Q)$ instead of $u_Q(P)$, for then we see that the result which we have just obtained can be rewritten in the form $K(P, Q) = K(Q, P)$. The equation appearing in Lemma 4 can now be written in the form

$$u(Q) = \iint_D u(P)K(P, Q)\, dx\, dy,$$

where the variables of integration x and y are the coordinates of P. From this equation it is apparent why $K(P, Q)$ is termed the *reproducing kernel* of the space $H(D)$.

Now let u be any function in $H(D)$ satisfying the condition $u(Q) = 1$, where Q is any fixed point of D. Then from the preceding results and the Schwarz inequality, we obtain

$$1 = u(Q) = (u, u_Q) = |(u, u_Q)| \leqslant \|u\| \cdot \|u_Q\|.$$

Therefore, $\|u\| \geqslant 1/\|u_Q\|$, and equality holds iff u is a multiple of u_Q; i.e., iff $u(P) \equiv cK(P, Q)$. Setting $P = Q$, we obtain $1 = cK(Q, Q)$, or $c = 1/K(Q, Q)$. Thus, of all members of $H(D)$ satisfying the condition $u(Q) = 1$, the function $K(P, Q)/K(Q, Q)$ possesses the smallest norm, which is

$$\frac{1}{\|u_Q\|} = \frac{1}{(u_Q, u_Q)^{1/2}} = \frac{1}{(u_Q(Q))^{1/2}} = \frac{1}{(K(Q, Q))^{1/2}}.$$

Finally, we proceed to obtain an expansion of the reproducing kernel $K(P, Q)$ in terms of an arbitrary orthonormal basis $\{v_1, v_2, \ldots, v_n, \ldots\}$. (We know that an orthonormal basis of $H(D)$ exists, consists of a countably infinite subset of $H(D)$, and that infinitely many choices of such a basis can be made.) For any function u in $H(D)$ we know that the expansion

$$u = \sum_{k=1}^{\infty} (u, v_k) v_k$$

holds, in the sense that $\|u - \sum_{k=1}^{n} (u, v_k)v_k\| \to 0$ with increasing n. If we denote the finite sum $\sum_{k=1}^{n} (u, v_k)v_k$ as u_n, then $\|u - u_n\| \to 0$, and by earlier results we conclude that the sequence $u_1(P), u_2(P), u_3(P), \ldots$ converges *pointwise* to $u(P)$ for all $P \in D$ and that the convergence is uniform on each compact subset of D. Thus we may write

$$u(P) = \sum_{k=1}^{\infty}(u, v_k)v_k(P)$$

with the assurance that the infinite series appearing on the right actually converges. If, in particular, we choose for u the function u_Q, we obtain

$$u_Q(P) = K(P, Q) = \sum_{k=1}^{\infty}(u_Q, v_k)v_k(P) = \sum_{k=1}^{\infty} v_k(Q)v_k(P).$$

This expansion (which explicitly demonstrates that the reproducing kernel is a symmetric function of its two arguments) is especially remarkable in that it holds for *every* choice of the orthonormal basis $v_1, v_2, v_3, \ldots$.

(The reader may find it instructive at this point to review the problem discussed in §4–4.)

Exercises

1. Prove Lemma 2.

2. Prove that the functions $\operatorname{Re}(x + iy)^n$, $n = 0, 1, 2, 3, \ldots$, are harmonic and linearly independent in any domain; since they are bounded in any *bounded* domain D, it follows that $H(D)$ is infinite-dimensional.

3. Let $Q_1, Q_2, \ldots, Q_n$ be distinct points in the bounded domain D. Show that the functions $u_{Q_1}, u_{Q_2}, \ldots, u_{Q_n}$ are linearly independent.

4. Let $Q_1, Q_2, \ldots, Q_n$ be distinct points in the bounded domain D and let the (real) constants $\alpha_1, \alpha_2, \ldots, \alpha_n$ be given. Show that the problem of minimizing $\|u\|$ (in $H(D)$) subject to the interpolation conditions $u(Q_k) = \alpha_k$, $1 \leqslant k \leqslant n$, possesses a unique solution.

5. Let D and $\tilde{D}$ be bounded plane domains, with reproducing kernels K and $\tilde{K}$, respectively. Suppose that $Q \in D \subseteq \tilde{D}$. Prove that $K(Q, Q) \geqslant \tilde{K}(Q, Q)$.

6. A sequence $g_1, g_2, g_3, \ldots$ of vectors in a Hilbert space H is said to *converge weakly* if for every vector f in H the sequence of scalars $(f, g_1), (f, g_2), (f, g_3), \ldots$ is convergent.

(*a*) Prove that a convergent sequence of vectors is weakly convergent.
(*b*) Prove that a necessary condition for a sequence of vectors to converge weakly is that their norms be bounded.
(*c*) Demonstrate a sequence of vectors which converges weakly but does not converge.

§4. THE DUAL SPACE OF $L^p(A)$

Theorem 3–1 may be stated, somewhat imprecisely, in the following form: If H is any Hilbert space, then $H = H^*$. We now proceed to show that, in a sense to be made quite precise, $(L^p(A))^* = L^q(A)$ if $1 \leqslant p < \infty$. (The reader may find it helpful to refer back to the definition of $L^p(A)$ and to the definition of N^*, where N is any normed linear space.)

Let p be fixed, $1 \leqslant p \leqslant +\infty$, let the corresponding conjugate number q be determined, and let g be any member of L^q. (As in Chapter 2, we frequently use the notations L^p, L^q instead of $L^q(A)$, $L^q(A)$.) According to the Hölder inequality, $|\int fg| \leqslant \|g\|_q \cdot \|f\|_p$ for every f in L^p. Since the equality $\int(\alpha f_1 + \beta f_2)g = \alpha \int f_1 g + \beta \int f_2 g$ holds for every pair of functions $\{f_1, f_2\}$ in L^p and every pair of scalars $\{\alpha, \beta\}$, we see that the equation $l_g(f) = \int fg$ defines a bounded linear functional on L^p; furthermore, by reference to Exercise 2–2 of Chapter 3 it is easily seen that $\|l_g\| = \|g\|_q$. Thus, each member g of L^q determines a member of $(L^p)^*$, and the correspondence thus defined is norm-preserving. In particular, l_g is the zero functional iff $\|g\|_q = 0$, which in turn is equivalent to the condition $g(x) = 0$ a.e. Thus, distinct members of L^q determine distinct members of $(L^p)^*$. Furthermore, the correspondence is a linear one; this simply means that if α and β are any scalars and if g_1 and g_2 are any members of L^q, then $l_{\alpha g_1 + \beta g_2} = \alpha l_{g_1} + \beta l_{g_2}$.

The preceding remarks obviously suggest the following question: Given a member l of $(L^p)^*$, does there exist a member g of L^q such that $l = l_g$? The remarkable answer to this question is in the affirmative for $l \leqslant p < +\infty$ but in the negative for $p = +\infty$. A complete justification of this assertion would require the development of a portion of the theory of Lebesgue integration that we have not presented in Chapter 2—namely, the determination of the precise conditions under which a function f, defined on some interval, shall be the indefinite integral of some (summable) function F and the answer to the converse question: If F is summable in an interval $[a, b]$ and if $f(x) = \int_a^x F$ for $x \in [a, b]$, in what sense does the familiar equality $f'(x) = F(x)$, easily established when F is continuous, carry over to the present case?

First we shall present, through Definition 1 and Theorem 1, the answers to the two previously stated problems, but we shall not give the proof. Then we shall indicate briefly how to settle the question posed at the beginning of the preceding paragraph.

Definition 1: *Let f be a (real-valued or complex-valued) function defined on an interval (open or closed, bounded or unbounded). The function f is said to be absolutely continuous if for every $\epsilon > 0$ there exists a $\delta > 0$ such that for every finite disjoint collection of intervals $\{(a_1, b_1), (a_2, b_2), \ldots, (a_n, b_n)\}$ with total length less than δ the inequality $\sum_{k=1}^{n} |f(b_k) - f(a_k)| < \epsilon$ holds. (Of course, if f is absolutely continuous on some interval it is also absolutely continuous on any subinterval.)*

Theorem 1: *If f is absolutely continuous on a bounded closed interval $[a, b]$, then the derivative f' exists almost everywhere. Furthermore, f' is summable on the interval, and $\int_a^x f' = f(x) - f(a)$ for every $x \in [a, b]$. Conversely, if $F \in L^1([a, b])$, the indefinite integral of F, defined for all $x \in [a, b]$ by the equation*

$$f(x) = \int_a^x F,$$

is absolutely continuous, and $f'(x) = F(x)$ almost everywhere in $[a, b]$.

Now we can return to the main problem of this section. For ease in exposition, we choose for A the interval $[0, 1]$ and confine attention to the real, rather than the complex, $L^p(A)$—that is to say, the members of $L^p(A)$ are (equivalence classes of) real-valued functions and $l(f)$ is, for each f in $L^p(A)$, a real number. (The extension of our results to any other measurable set A and to the complex $L^p(A)$ is a task of very minor difficulty.) For each number a in A, let h_a be the characteristic function of the interval $[0, a]$, and let $H(a) = l(h_a)$. (Note carefully that there is a function h_a for each a, while H is just one function.) More generally, let $h_{a,b}$ be the characteristic function of the interval $[a, b]$, where it is always understood that $0 \leqslant a < b \leqslant 1$. Note that the equality $h_{a,b}(x) = h_b(x) - h_a(x)$ holds almost everywhere (in fact, everywhere except at a), so that $l(h_{a,b}) = l(h_b) - l(h_a) = H(b) - H(a)$.

The continuity of the function H is immediate: For $b > a$,

$$|H(b) - H(a)| = |l(h_{a,b})| \leqslant \|l\| \cdot \left\{\int_a^b 1^p\right\}^{1/p} = \|l\| \cdot (b - a)^{1/p},$$

which approaches zero as $b - a$ does so. In order to show that H is, in fact, *absolutely* continuous, select any finite collection of disjoint intervals (it really makes no difference whether they are open, closed, or half-open), $\{(a_1, b_1), (a_2, b_2), \ldots, (a_n, b_n)\}$. Define a function f as follows: $f = +1$ everywhere in the interval (a_k, b_k) if $H(b_k) - H(a_k) \geqslant 0$, $f = -1$ everywhere in the interval (a_k, b_k) if $H(b_k) - H(a_k) < 0$, and $f = 0$ outside the union of the given collection of intervals. Obviously $f \in L^p$, and

$$\|f\|_p = \left\{\sum_{k=1}^{n} \int_{a_k}^{b_k} 1^p\right\}^{1/p} = \left\{\sum_{k=1}^{n} (b_k - a_k)\right\}^{1/p};$$

furthermore,

$$l(f) = l\left(\sum_{k=1}^{n} \pm (h_{b_k} - h_{a_k})\right) = \sum_{k=1}^{n} \pm (l(h_{b_k}) - l(h_{a_k})) = \sum_{k=1}^{n} |H(b_k) - H(a_k)|.$$

But $|l(f)| \leqslant \|l\| \cdot \|f\|_p$, and so

$$\sum_{k=1}^{n} |H(b_k) - H(a_k)| \leqslant \|l\| \cdot \left\{\sum_{k=1}^{n} (b_k - a_k)\right\}^{1/p}.$$

Given any $\epsilon > 0$, let $\delta = (\epsilon/\|l\|)^p$. Then whenever $\sum_{k=1}^{n} (b_k - a_k)$ is less than δ, the quantity $\|l\| \cdot \{\sum_{k=1}^{n} (b_k - a_k)\}^{1/p}$ is less than $\|l\| \cdot \delta^{1/p}$, which is exactly ϵ. Thus, $\sum_{k=1}^{n} |H(b_k) - H(a_k)| < \epsilon$, and so H is absolutely continuous. (Note carefully that the disjointness of the chosen set of intervals and the finiteness of p are both essential in the preceding argument.)

Now, by Theorem 1, H' exists almost everywhere, and $H(b) - H(a) = \int_a^b H'$. If f is the characteristic function of the interval (a, b), the equalities $l(f) = H(b) - H(a) = \int_a^b H' = \int_a^b fH' = \int_0^1 fH'$ must hold. (The last equality follows from the fact that f vanishes outside (a, b).) If E is the union of a finite collection of open intervals, then by an obvious extension of the preceding argument we obtain the equality

$$l(\chi_E) = \int_0^1 \chi_E H'. \tag{4–1}$$

(Recall that $\chi_E(x)$ equals 1 or 0, according as x does or does not belong to E.) By an easy passage to the limit we find that (4–1) continues to hold if E is the union of a countably infinite collection of open intervals; hence, (4–1) holds for all open subsets of A (= [0, 1]). Using the fact that every measurable set is almost open (in the sense made precise by the definition of measurability), we can establish (4–1) whenever E is measurable. Then by taking (finite) linear combinations of characteristic functions we obtain, for any simple function* s, the obvious generalization of (4–1):

$$l(s) = \int_0^1 sH'. \tag{4–2}$$

---

* Here, in contrast to Chapter 2, we may obviously permit s to assume negative values.

By an easy limiting operation we then obtain, for any bounded measurable function f (which clearly belongs to $L^p(A)$ for all values of p, including $+\infty$), the formula

$$l(f) = \int_0^1 fH'. \tag{4–3}$$

Finally, by a more delicate limiting operation we can show that (4–3) continues to hold for the unbounded members of $L^p(A)$.

Replacing H' by g, we obtain the desired representation of the bounded linear functional l, namely

$$l(f) = \int_0^1 fg. \tag{4–4}$$

However, it still remains to show that $g \in L^q(A)$; Theorem 1 guarantees only that $g \in L^1(A)$. We present this argument in some detail, since it provides a splendid example of the power of the principle of uniform boundedness. Let the sequence of functions $g_1, g_2, g_3, \ldots$ be defined as the *truncates* of g: $g_n(x) = -n$, $g(x)$, or n according as $g(x) < -n$, $|g(x)| \leqslant n$, or $g(x) > n$, and for each f in $L^p(A)$ let

$$l_n(f) = \int_0^1 fg_n. \tag{4–5}$$

Each l_n is a bounded linear functional, for, by the Hölder inequality, $|l_n(f)| \leqslant \|f\|_p \, \|g_n\|_q \leqslant n \, \|f\|_p$. Now, by splitting A into the disjoint sets A_+ and A_- on which $fg \geqslant 0$ and $fg < 0$, respectively, we can easily show, by invoking the monotone convergence theorem, that $\lim_{n\to\infty} \int_0^1 fg_n$ exists and equals $\int_0^1 fg$. Thus, the collection of numbers $\{l_n(f)\}$ is bounded at each point of $L^p(A)$, and since $L^p(A)$ is complete we are now assured that the set of norms $\{\|l_n\|\}$ is bounded. Since $\|l_n\|$ is known to equal $\|g_n\|_q$, we have shown that the set of norms $\{\|g_n\|_q\}$ is bounded. By another application of the monotone convergence theorem (where this time we split A into the subsets on which $g \geqslant 0$ and $g < 0$) we conclude that $\|g\|_q = \lim_{n\to\infty} \|g_n\|_q < +\infty$. Thus, we have sketched the proof of the following remarkable theorem.

Theorem 2: *For $1 \leqslant p < +\infty$, $(L^p)^* = L^q$, where the equality is to be understood in the manner explained in the third paragraph of this section.*

For a fully detailed proof of Theorems 1 and 2, the reader is referred to [Royden, pp. 94–107 and 119–123].

We have thus shown (except for the gaps in the argument that have been indicated) that there exists a one-to-one correspondence between

$(L^p)^*$ and L^q which preserves both the linear structure and the metric structure of these two spaces, so that, when looked upon simply as normed linear spaces, without regard for the nature of the vectors which constitute the two spaces, they are indistinguishable.

The case $p = 2$ has already been covered by Theorem 3–1, for L^2 becomes a Hilbert space with the definition $(f, g) = \int f\bar{g}$ for the inner product. A slight confusion may be caused by the presence of $\bar{g}$, rather than g. This is readily cleared up by observing that if, according to Theorem 3–1, the bounded linear functional l is associated with the vector g, then the bounded linear functional αl is associated with the vector $\bar{\alpha}g$, not with the vector αg. Thus, strictly speaking, Theorem 3–1 furnishes a *conjugate isomorphism*, rather than an isomorphism, between L^2 and $(L^2)^*$. (Of course, this complication does not arise if we are dealing with real-valued, rather than complex-valued, functions.)

We leave to the reader, as Exercise 2, the task of proving the following analogue of Theorem 2.

Theorem 3: *For* $1 \leqslant p < +\infty$, $(l^p)^* = l^q$.

(As might be expected, the proof of this theorem is considerably simpler than that of Theorem 2, since no measure theory is involved.)

Exercises

1. Demonstrate a real-valued function which is uniformly, but not absolutely, continuous on the interval $[0, 1]$.
2. Prove Theorem 3.

§5. THE HAHN-BANACH THEOREM

If N is any normed linear space, the dual space N^* is not empty, for it certainly contains the zero functional. It appears almost obvious that (aside from the trivial case when N consists exclusively of the zero vector) N^* must contain other functionals as well. Incidentally, Theorem 1–2 gives a very complete picture of the structure of N^* when N is finite-dimensional; the absence of any reference in Theorem 1–2 to a norm is readily remedied by norming N in any manner consistent with the definition of a norm. The case of an infinite-dimensional normed linear space is more intricate and involves the use of a logical principle, namely Zorn's lemma, which plays a rather peculiar and controversial role in the development of mathematics. However, we shall use it unquestioningly at the decisive point in the development of the present section, but we shall provide an exceedingly brief explanation of this principle for the benefit of the reader who may not be acquainted with it.

Definition 1: *Let L be any linear space, let M be any non-empty subset of L, and let f be any vector in L. By* $[f] + M$ *we mean the collection of all vectors of the form* $\alpha f + g$*, where* α *is a scalar and* $g \in M$.

Theorem 1: *Let* $\mathcal{M}$ *be a linear manifold of a linear space L.*

(*a*) *If f is any vector in L, then* $[f] + \mathcal{M}$ *is also a linear manifold; if* $f \in \mathcal{M}$*, then* $[f] + \mathcal{M} = \mathcal{M}$*, while if* $f \notin \mathcal{M}$*, then* $\mathcal{M}$ *is a proper subset of* $[f] + \mathcal{M}$.

(*b*) *If S is a subspace of a normed linear space N and if f is any vector in N, then* $[f] + S$ *is also a subspace.*

PROOF: (*a*) Left to reader as Exercise 5.

(*b*) If $f \in S$, then, by part (*a*), $[f] + S = S$, and so $[f] + S$ is certainly a subspace. Therefore, we may confine attention to the case when $f \notin S$. We know from (*a*) that $[f] + S$ is a linear manifold, so we have only to prove that $[f] + S$ is closed. Suppose not—then there would exist a limit-point of $[f] + S$ which does not belong to $[f] + S$. Calling this vector h, we could find a sequence $h_1, h_2, h_3, \ldots$ of vectors which *do* belong to $[f] + S$ and which converge to h. Each vector h_n can be written in the form $\alpha_n f + g_n$, where $g_n \in S$. There are two possibilities—the set of numbers $\{\alpha_n\}$ is bounded or it is unbounded. Suppose it is unbounded. Then by confining attention to a suitably chosen subsequence, if necessary, we may suppose that $|\alpha_n| \to \infty$. Each g_n may be expressed as $-\alpha_n \tilde{g}_n$, where $\tilde{g}_n$ also belongs to S. Then, dividing the limit relation $\alpha_n f + g_n \to h$ by α_n, we obtain $f - \tilde{g}_n \to o$ (since $(1/\alpha_n)h \to o$); but this says that f is a limit-point of S, contradicting the hypothesis that S is closed.

Therefore, the set of numbers $\{\alpha_n\}$ is bounded, and by the Bolzano-Weierstrass property we can select a subsequence of the α_n's which converges to a number α. Then $\alpha_n f \to \alpha f$, and from the relation $\alpha_n f + g_n \to h$, we see that g_n must approach a limit, which we call g. Thus, $\alpha f + g = h$. Also, since the g_n's all belong to S and since S is closed, g must belong to S, and so $h \in [f] + S$. Thus $[f] + S$ contains all its limit-points—i.e., it is closed.

Theorem 2: *Let S be a proper subspace of the real normed linear space N, let l be a bounded linear functional on S* (*not on N*), *and let f be a vector belonging to* $N - S$. *Then it is possible to extend l from S to* $[f] + S$ *without increasing its norm.* (*That is, there exists a bounded linear functional on* $[f] + S$ *whose norm is equal to* $\|l\|$ *and whose restriction to S is l.*)

PROOF: (*a*) Every vector belonging to $[f] + S$ is expressible in the form $\alpha f + g$, where $\alpha \in R$ and $g \in S$; the vector g and the number α are *uniquely* determined by the given vector, for if there were two *different* representations we would have $\alpha_1 f + g_1 = \alpha_2 f + g_2$, or $(\alpha_1 - \alpha_2)f = g_2 - g_1$, or $f = (1/(\alpha_1 - \alpha_2))(g_2 - g_1)$. This says that f certainly belongs to S, contrary to assumption. Thus, $\alpha_1 = \alpha_2$, and so $g_1 = g_2$, and we have the uniqueness. Obviously, the number α is zero iff the vector under consideration belongs to S.

(*b*) Suppose that the desired extension of l exists; call it $\tilde{l}$. Then for any vector h in $[f] + S$ we have $h = \alpha f + g$, where the scalar α and the vector $g \in S$ are uniquely determined by h, and so we must have $\tilde{l}(h) = \alpha\tilde{l}(f) + \tilde{l}(g)$. However, since $\tilde{l}$ is an extension of l, this last equation reduces to $\tilde{l}(h) = \alpha\tilde{l}(f) + l(g)$, and so we see that the problem of determining $\tilde{l}$ reduces to the problem of determining suitably *one fixed* number, $\tilde{l}(f)$. Now, it is clear that no matter *what* value we assign to $\tilde{l}(f)$, say β, the equation $\tilde{l}(h) = \alpha\beta + l(g)$ determines a *linear* functional on $[f] + S$ which extends l, but is it not necessarily true that $\tilde{l}$ will have the same norm as l; for example, if we choose $\beta = 1 + \|l\| \cdot \|f\|$ we clearly have $|\tilde{l}(f)| = 1 + \|l\| \cdot \|f\| > \|l\| \cdot \|f\|$, and so $\|\tilde{l}\| > \|l\|$. (In fact, it is not even obvious, but it happens to be true, that *every* choice of β makes $\tilde{l}$ a bounded linear functional.) Thus, our problem is to show that there exists *at least* one choice of the number β which will *prevent* $\|\tilde{l}\|$ from exceeding $\|l\|$. (Of course, it is impossible to accomplish $\|\tilde{l}\| < \|l\|$.)

(*c*) If l is the zero functional, there is nothing to prove—simply choose $\beta = 0$. Next, suppose that $\|l\| = 1$. Then we have to show that there exists a number β such that for every real number α and for every vector $g \in S$ the inequality

$$|\alpha\beta + l(g)| \leqslant \|\alpha f + g\|$$

must hold. We consider two separate cases—$\alpha = 0$, $\alpha \neq 0$.

(*i*) If $\alpha = 0$, any β will suffice, since the inequality becomes $|l(g)| \leqslant \|g\|$, which is certainly true, since $\|l\| = 1$.

(*ii*) If $\alpha \neq 0$, the preceding inequality is equivalent to

$$\left|\beta + l\left(\frac{g}{\alpha}\right)\right| \leqslant \left\|f + \frac{g}{\alpha}\right\|.$$

Now, for any non-zero α, g/α sweeps out S as g sweeps out S. Hence, we have succeeded in eliminating α from our problem—we have the following simplified problem: To show that it is possible to choose β in such a way that $|\beta + l(g)| \leqslant \|f + g\|$ for *every* g in S. Now, choose any two vectors, g_1 and g_2, in S. Since l is defined and linear on S and since $\|l\| = 1$, we certainly have

$$\begin{aligned} l(g_1) - l(g_2) = l(g_1 - g_2) \leqslant \|g_1 - g_2\| &= \|(g_1 + f) + (-g_2 - f)\| \\ &\leqslant \|g_1 + f\| + \|-g_2 - f\| = \|f + g_1\| + \|f + g_2\|, \end{aligned}$$

and so

$$-l(g_2) - \|f + g_2\| \leqslant -l(g_1) + \|f + g_1\|.$$

Now, *fix* g_2, and let g_1 vary freely over S. The right side cannot go below the left side, and so we conclude that the right side is *bounded below*, and that $-l(g_2) - \|f + g_2\| \leqslant \inf_{g\in S} \{-l(g) + \|f + g\|\}$ for *any* $g_2 \in S$. Now let g_2 vary freely over S. Repeating the preceding argument, we obtain

$$\sup_{g\in S} \{-l(g) - \|f + g\|\} \leqslant \inf_{g\in S} \{-l(g) + \|f + g\|\}.$$

For convenience we denote the left and right sides of this last inequality as γ and δ, respectively. Let β be chosen as *any* number in the closed interval $[\gamma, \delta]$ (which may consist of a single point, but, as we have seen, is certainly not empty). With this choice of β we have, for any vector $g \in S$,

$$-l(g) - \|f + g\| \leqslant \beta \leqslant -l(g) + \|f + g\|,$$

or

$$-\|f + g\| \leqslant \beta + l(g) \leqslant \|f + g\|,$$

or

$$|\beta + l(g)| \leqslant \|f + g\|,$$

which is exactly what we wanted.

(*d*) Finally, if $\|l\| \neq 0$ and $\|l\| \neq 1$, simply apply the previous reasoning to the bounded linear functional $(1/\|l\|)l$, which has norm 1.

Remark: Theorem 2 has been stated only for a normed linear space over the real field, and the proof depends very heavily on the ordering of the real numbers by the relation $\leqslant$. The theorem itself is, in fact, true for a complex normed linear space also, but, since the present theorem is only a preliminary to Theorem 5, we merely point out here that the proof of the latter theorem will indicate clearly how Theorem 2 can be extended from the real to the complex field of scalars.

Theorem 3: *Let N be a normed linear space, let $\mathcal{M}$ be a linear manifold, but not a subspace, of N. Then $\overline{\mathcal{M}}$ is a subspace of N.*

PROOF: Trivial.

Theorem 4: *Let N be a normed linear space, let $\mathcal{M}$ be a linear manifold, but not a subspace, of N, and let l be a bounded linear functional defined only on $\mathcal{M}$. Then l can be extended in a unique manner to become a bounded linear functional on $\overline{\mathcal{M}}$, and the extension has exactly the same norm as l.*

PROOF: Trivial.

Theorem 5 (Hahn-Banach): *Let N be a normed linear space, real or complex, let S be a proper subspace of N, and let l be a bounded linear functional on S. Then l can be extended to N without increasing the norm.**

PROOF: It appears to be in the nature of things that this theorem must first be proven for the real field, after which the case of the complex field is treated as a corollary. Accordingly, we divide the proof into two portions.

(*a*) If N is a *real* normed linear space, Theorem 2 assures us that there exists a *proper* extension of l—i.e., a bounded linear functional $\tilde{l}$, defined on a subspace $\tilde{S}$ which properly contains S, such that $\|\tilde{l}\| = \|l\|$ and l is the restriction of $\tilde{l}$ to S. In other words, the collection P of all norm-preserving extensions of l to linear manifolds properly containing S is non-empty. If l_1 and l_2 are members of P such that l_2 is either the same as l_1 or a proper extension of l_1, we write $l_1 \leqslant l_2$. The relation $\leqslant$ obviously constitutes a partial ordering of P; furthermore, if C is any chain in P (with respect to this partial ordering), the chain C determines unambiguously a particular member l_C of P in the following way: For any g appearing in the domain of any member $\tilde{l}$ of C, we define $l_C(g)$ to be $\tilde{l}(g)$. (Note carefully that, since C is a *chain*, two different members of C having g in their domains must agree at g, so that, as asserted previously, $l_C(g)$ is indeed *unambiguously* defined.) It is not difficult to see that l_C is a linear functional, and since $|l_C(g)| = |\tilde{l}(g)| \leqslant \|\tilde{l}\| \cdot \|g\| = \|l\| \cdot \|g\|$, it follows that $\|l_C\| = \|l\|$; furthermore, $l_C \geqslant \tilde{l}$ for every member $\tilde{l}$ of the chain C. According to Zorn's lemma, there exists a member of P, which we shall denote l_m (the letter m suggesting "maximal"), having the property that there does not exist in P any *proper* extension of l_m. By referring to Theorem 4, we see that the domain of l_m must be a subspace, and then by referring to Theorem 2 we see that the domain of l_m must coincide with N. Thus, the proof is complete in the case of a real normed linear space.

(*b*) If N is a *complex* normed linear space, we can also consider N as a *real* normed linear space by the simple device of restricting scalar multipliers of vectors to the real field. We shall find it convenient to refer to N, when considered as a *real* normed linear space, as N_r. For any vector f in the domain of l, we may write $l(f) = g(f) + ih(f)$, where g and h are real-valued functionals. If α is real, we immediately obtain, on the one hand, $l(\alpha f) = g(\alpha f) + ih(\alpha f)$, and, on the other hand, $l(\alpha f) = \alpha l(f) = \alpha g(f) + i\alpha h(f)$. Thus, g and h are (real) linear functionals. On the other hand, $l(if) = g(if) + ih(if) = il(f) = -h(f) + ig(f)$.

* If the reader is not acquainted with Zorn's lemma, he should at this time turn to Appendix *A* and then return to the proof of the present theorem.

Hence, $h(f) = -g(if)$, and so we may express the *complex* linear functional l entirely in terms of the *real* linear functional g as follows: $l(f) = g(f) - ig(if)$. (We dispense with the trivial proof that the functional g is actually linear.) Since $|g(f)| = |\text{Re}\, l(f)| \leqslant |l(f)| \leqslant \|l\| \cdot \|f\|$, we see that g is a bounded linear functional defined on the subspace S of N_r. By the first half of this proof, we can extend g to a bounded linear functional g_m defined on all of N_r without increasing the norm. Then it is readily confirmed that the functional l_m, defined on all of N by the relation

$$l_m(f) = g_m(f) - ig_m(if),$$

is indeed a linear functional. Furthermore, $|l_m(f)| \leqslant |g_m(f)| + |g_m(if)| \leqslant \|g_m\| \cdot (\|f\| + \|if\|) = 2\,\|g_m\| \cdot \|f\| \leqslant 2\,\|l\| \cdot \|f\|$, and so $\|l_m\| \leqslant 2\,\|l\|$. In order to eliminate the factor 2, we employ the following trick: For any $f \in N$, we can find a complex number α, $|\alpha| = 1$, such that $(1/\alpha)l_m(f)$ is real and non-negative. Then $l_m((1/\alpha)f) = g_m((1/\alpha)f) - ig_m((i/\alpha)f) = g_m((1/\alpha)f)$ and therefore $|l_m(f)| = |(1/\alpha)l_m(f)| = |l_m((1/\alpha)f)| = |g_m((1/\alpha)f)| \leqslant \|g_m\| \cdot \|(1/\alpha)f\| = \|g_m\| \cdot \|f\| \leqslant \|l\| \cdot \|f\|$, and so $\|l_m\| \leqslant \|l\|$. Since l_m is an extension of l (to all of N), the inequality $\|l_m\| < \|l\|$ cannot hold, and so $\|l_m\| = \|l\|$.

Corollary: *(a) If f is any non-zero vector in a normed linear space N, there exists a member l of N^* such that $l(f) = \|f\|$ and $\|l\| = 1$.*

(b) N^* separates N, in the sense that, given any two distinct vectors g, h of N, there exists a member l of N^* such that $l(g) \neq l(h)$.

PROOF: (*a*) Let S be the subspace of N consisting of all vectors of the form αf, and for each vector g $(= \alpha f)$ in S let $l(g) = \alpha\,\|f\|$. Clearly, $l(f) = \|f\|$ and $\|l\| = 1$. By Theorem 5, we can extend l to all of N without increasing its norm.

(*b*) Let $f = g - h$ and choose l in accordance with part (*a*). Then $l(g) - l(h) = l(g - h) = l(f) = \|f\| \neq 0$, so that $l(g) \neq l(h)$.

We conclude by noting that the completeness of N is not assumed anywhere in this section; however, the completeness of the field of scalars (R or $\mathfrak{C}$) plays an essential role.

Exercises

1. Prove part (*a*) of Theorem 1.
2. Prove the assertion made in the first parenthetical remark appearing in part (*b*) of the proof of Theorem 2.

CHAPTER 6

OPERATORS

§1. LINEAR TRANSFORMATIONS AND OPERATORS

Although we can consider mappings from one linear space into another (we have done this when we developed the theory of linear functionals in the preceding chapter), we shall confine attention in this chapter entirely to mappings from a given linear space into itself. For definiteness we consider complex spaces.

Definition 1: *A linear transformation is a mapping T of a linear space V into V satisfying the condition $T(\alpha f + \beta g) = \alpha(T(f)) + \beta(T(g))$ for all vectors f, g and all scalars α, β. (Clearly, $T(o) = o$.) We write Tf instead of $T(f)$ whenever no confusion can result.*

Definition 2: *If α is a scalar and T is a linear transformation, then $\alpha \cdot T$ is the mapping S defined in the obvious manner: $Sf = \alpha(Tf)$ for all vectors f. The sum $T_1 + T_2$ of two linear transformations is the mapping S defined in the obvious manner: $Sf = (T_1f) + (T_2f)$ for all vectors f. (The mapping $\alpha \cdot T$ is usually denoted αT.)*

Examples

(*a*) Let V be the class of ordered pairs of scalars and let the vector a be the ordered pair (α_1, α_2). The mapping T defined by the equation $Ta = (\gamma_1\alpha_1 + \gamma_2\alpha_2, \gamma_3\alpha_1 + \gamma_4\alpha_2)$, where $\gamma_1, \gamma_2, \gamma_3, \gamma_4$ are four arbitrary, but fixed, scalars, is clearly a linear transformation. The mapping αT is obviously obtained by replacing $\gamma_1, \gamma_2, \gamma_3, \gamma_4$ by $\alpha\gamma_1, \alpha\gamma_2, \alpha\gamma_3, \alpha\gamma_4$ respectively. If, similarly, $\tilde{T}$ is the linear transformation obtained by employing scalars $\tilde{\gamma}_1, \tilde{\gamma}_2, \tilde{\gamma}_3, \tilde{\gamma}_4$, then $T + \tilde{T}$ is the linear transformation obtained by employing the scalars $\gamma_1 + \tilde{\gamma}_1, \gamma_2 + \tilde{\gamma}_2, \gamma_3 + \tilde{\gamma}_3, \gamma_4 + \tilde{\gamma}_4$.

(*b*) Let V be the class of continuous complex-valued functions defined on the interval $[0, 1]$, and for each vector f let Tf be the function g defined by the equation $g(x) = \int_0^x f(t)\, dt$.

Note that in example (*a*) the linear transformation T is an *onto* mapping* iff $\gamma_1\gamma_4 - \gamma_2\gamma_3 \neq 0$, while in example (*b*) the linear transformation T is certainly *not* onto (why?).

Theorem 1: *The class of all linear transformations on V, denoted $L(V)$, is itself a linear space (with the definitions of $\cdot$ and $+$ given in Definition 2).*

PROOF: Trivial.

Definition 3: *The product T_1T_2 of the linear transformations T_1 and T_2 is the transformation S defined in the obvious manner: $Sf = T_1(T_2f)$ for all vectors f.*

Theorem 2: *The product of two linear transformations is again a linear transformation. In general, $T_1T_2 \neq T_2T_1$; if equality holds, then T_1 and T_2 are said to commute.*

PROOF: Left as Exercise 1.

Theorem 3: *If T_1, T_2, T_3 are any three linear transformations, then $T_1(T_2T_3) = (T_1T_2)T_3$, so that we may simply write $T_1T_2T_3$. (By induction it follows that $T_1T_2 \cdots T_n$ is unambiguously defined.)*

PROOF: Left as Exercise 2.

Definition 4: *The linear transformations O (zero) and I (identity) are defined in the obvious manner: $Of = o$ and $If = f$ for every vector f. Obviously $OT = TO = O$ and $IT = TI = T$ for every linear transformation T.*

Theorem 4: *For any linear transformations T_1, T_2, T_3 and scalars α, β, the equality $T_1(\alpha T_2 + \beta T_3) = \alpha(T_1T_2) + \beta(T_1T_3)$ holds.*†

---

* Recall that a mapping from a set A to a set B is said to be *onto* if each member of B is the image of at least one member of A—that is, if the range of the mapping coincides with B.

† Theorems 1 through 4 can be condensed into the single statement that $L(V)$ is an *algebra*. We shall not present a precise definition of this term, however; roughly, it denotes a linear space in which any pair of vectors (as well as a scalar and a vector) can be multiplied.

PROOF: Left as Exercise 3.

Now we turn to the case that V is a normed linear space; we therefore denote the space as N rather than V. The ideas presented here constitute an obvious generalization of some of the ideas concerning bounded linear functionals.

Definition 5: *A linear transformation T defined on N is said to be bounded if there exists a real number C such that $\|Tf\| \leqslant C\,\|f\|$ for every vector f. If T is bounded, $\|T\|$, called the norm of T, is defined as the* inf *of all values of C for which the preceding inequality holds for all vectors f. (Cf. Exercises* 4, 5, 6.) *A bounded linear transformation is henceforth called an operator.*

Theorem 5: *(a) $\|\alpha T\| = |\alpha| \cdot \|T\|$ for any scalar α and operator T.*
(b) $\|T_1 + T_2\| \leqslant \|T_1\| + \|T_2\|$ for any operators T_1 and T_2.
(c) $\|T_1T_2\| \leqslant \|T_1\| \cdot \|T_2\|$ for any operators T_1 and T_2.
*(d)** $\|O\| = 0$, $\|I\| = 1$.

PROOF: Left as Exercise 7.

Definition 6: *For any operator T and positive integer n, we define T^n as $\underbrace{TT\cdots T}_{n \text{ times}}$. We occasionally find it convenient to define T^0 as I.*

Definition 7: *A non-zero operator T is said to be nilpotent if $T^n = O$ for some positive integer n. (Cf. Exercise* 8.)

Exercises

1. Prove Theorem 2.
2. Prove Theorem 3.
3. Prove Theorem 4.
4. Give an example of an unbounded linear transformation.
5. Show that a linear transformation defined on a finite-dimensional normed linear space is certainly bounded.
6. Show that, if T is an operator, then
$$\|T\| = \sup_{\|f\|=1} \|Tf\| = \sup_{\|f\|\leqslant 1} \|Tf\| = \sup_{f\neq o} \frac{\|Tf\|}{\|f\|}\,.$$
7. Prove Theorem 5.
8. Give an example of a nilpotent operator.

* Strictly speaking, the second assertion is correct only if N contains a non-zero vector, for when N consists exclusively of the vector o (and only in this case), $I = O$.

§2. THE ADJOINT OPERATOR

We shall now show that any operator T on a normed linear space N determines in a natural manner an *associated*, or *dual*, operator on N^*. Let l denote any bounded linear functional on N (i.e., $l \in N^*$). Then for each $f \in N$ let us define $l_T(f)$ as follows: $l_T(f) = l(Tf)$. Clearly, for any vectors f, g and scalars α, β the following chain of equalities hold: $l_T(\alpha f + \beta g) = l(T(\alpha f + \beta g)) = l(\alpha(Tf) + \beta(Tg)) = \alpha l(Tf) + \beta l(Tg) = \alpha l_T(f) + \beta l_T(g)$. Thus, l_T is a linear functional on N. Furthermore, $|l_T(f)| = |l(Tf)| \leqslant \|l\| \cdot \|Tf\| \leqslant (\|l\| \cdot \|T\|) \|f\|$, and so l_T is bounded; in fact, $\|l_T\| \leqslant \|T\| \cdot \|l\|$. Also, it is obvious that if l and $\tilde{l}$ are members of N^* and if α and β are any scalars, then $(\alpha l + \beta \tilde{l})_T = \alpha l_T + \beta \tilde{l}_T$. The results of the last two sentences can be summed up by saying that the mapping of N^* into N^* defined by $l \to l_T$ is an operator. We denote this operator T^* and call it the *adjoint* of T. Note carefully that, while T is an operator on N, T^* is an operator on N^*.

Of course, one can now consider $(T^*)^*$, which is an operator on N^{**}. We have seen in Chapter 4 that $N \subseteq N^{**}$, so that $(T^*)^*(f)$ is well-defined for every $f \in N$. As one might expect, $(T^*)^*(f) = Tf$ (cf. Exercise 1), and so $(T^*)^*$ is an extension of T. (In general, $(T^*)^*$ is a *proper* extension of T, for N is usually a *proper* subset of N^{**}.)

Theorem 1: (*a*) $(T_1 + T_2)^* = T_1^* + T_2^*$;
(*b*) $(\alpha T)^* = \alpha T^*$;
(*c*) $(I)^* = I^*$;
(*d*) $(T_1 T_2)^* = T_2^* T_1^*$.

PROOF: Trivial. Of course, (*c*) means that the identity on N has as its adjoint the identity on N^*.

Theorem 2: $\|T^*\| = \|T\|$.

PROOF: The inequality $\|l_T\| \leqslant \|T\| \cdot \|l\|$, which was established in the first paragraph of this section, shows that $\|T^*\| \leqslant \|T\|$. As for the reverse inequality, it is obviously true if $\|T\| = 0$. Therefore we may confine attention to the case $\|T\| > 0$, and by homogeneity it suffices to prove that $\|T^*\| \geqslant 1$ when $\|T\| = 1$. Given any positive number $\alpha < 1$ (say $\alpha = 0.999$), we can find a unit vector f such that $\|Tf\| > \alpha$. Let Tf temporarily be called g. By the corollary to the Hahn-Banach theorem, there exists a linear functional l such that $\|l\| = 1$ and $|l(g)| = \|g\|$. Thus, $\alpha = \|g\| = |l(Tf)| = |(T^*l)(f)| \leqslant \|T^*l\| \cdot \|f\| = \|T^*l\| \leqslant \|T^*\| \cdot \|l\| = \|T^*\|$. Hence, $\|T^*\| \geqslant \alpha$, and since α can be chosen arbitrarily close to 1, we obtain the desired inequality, $\|T^*\| \geqslant 1$. This completes the proof.

When we take account of Theorem 3–1 of Chapter 5, we see that the preceding discussion of the adjoint operator assumes an especially simple and elegant form in a Hilbert space H. Every vector g in H determines a member of H^*, and every member of H^* is determined by a (unique) member of H. Furthermore, if for clarity we denote the bounded linear functional on H determined by g by the symbol l_g, we have $l_{g+h} = l_g + l_h$ and $\|l_g\| = \|g\|$. Thus, the correspondence $g \leftrightarrow l_g$ between H and H^* is additive and norm-preserving. However, a small complication arises when we observe the effect of multiplying g by a scalar α:

$$l_{\alpha g}(f) = (f, \alpha g) = \bar{\alpha}(f, g) = \bar{\alpha} l_g(f).$$

Thus, instead of obtaining $l_{\alpha g} = \alpha l_g$, we have $l_{\alpha g} = \bar{\alpha} l_g$. Therefore, the mapping $g \to l_g$ is not linear; it is *anti-linear*, or *conjugate-linear*. (This complication does not arise, of course, when we deal with a real Hilbert space.)

Since H and H^* are essentially the same spaces, it is rather natural to think of T^* as an operator defined on H rather than on H^*. To be more precise, we reason in the following manner, which enables us to disregard completely the concept of the dual space N^* of a normed linear space N: Let T be an operator on H and let g be a fixed vector in H, while the vector f varies over all of H. Then the expression (Tf, g) obviously defines a linear functional on H; furthermore, this functional is bounded, for $|(Tf, g)| \leqslant \|g\| \cdot \|Tf\| \leqslant (\|g\| \cdot \|T\|)\, \|f\|$. By the Riesz representation theorem, there exists a unique vector g_T such that $(Tf, g) = (f, g_T)$ for all f. Now let us consider how g_T behaves as g varies freely over H. Clearly, $(Tf, g + h) = (Tf, g) + (Tf, h) = (f, g_T) + (f, h_T) = (f, g_T + h_T)$, and so $(g + h)_T = g_T + h_T$. Also, $(Tf, \alpha g) = \bar{\alpha}(Tf, g) = \bar{\alpha}(f, g_T) = (f, \alpha g_T)$, and so $(\alpha g)_T = \alpha g_T$. Hence, the mapping $g \to g_T$ is linear (*not* anti-linear). We now proceed to show that this mapping is also bounded—i.e., that there exists a number C such that $\|g_T\| \leqslant C\, \|g\|$ for all vectors g. We simply replace f in the equality $(Tf, g) = (f, g_T)$ by g_T, and so we obtain $(Tg_T, g) = (g_T, g_T)$; by the Schwarz inequality we obtain $\|g_T\|^2 \leqslant \|Tg_T\| \cdot \|g\| \leqslant \|T\| \cdot \|g_T\| \cdot \|g\|$, and then by dividing by $\|g_T\|$ we obtain $\|g_T\| \leqslant \|T\| \cdot \|g\|$. (If $g_T = o$ the division is not legitimate, but the last inequality is trivially true in this case.) Hence, the linear mapping $g \to g_T$ has been shown to be bounded, with norm not exceeding $\|T\|$.

The mapping $g \to g_T$ is henceforth denoted T^*, and called the adjoint of T. This is not quite correct, for T^* should, according to our earlier definition, be an operator on H^*, which is closely related to, *but not isomorphic to*, H (because of the unpleasant appearance of the conjugate in the relation $\bar{\alpha}(f, g_T) = (f, \alpha g_T)$). Nevertheless, when working in Hilbert spaces we consider T^* to operate on H, not on H^*. (Of course, this *cannot* be done on an *arbitrary* normed linear space.)

Thus, if T is an operator on H, T^* is the operator on H defined by the condition: $(Tf, g) = (f, T^*g)$ for all vectors f and g. We have already shown that $\|T^*\| \leqslant \|T\|$; however, it is almost trivial (cf. Exercise 2) that the adjoint of T^* (which certainly exists) must coincide with T, and so, upon replacing T in the last inequality by T^*, we obtain $\|T\| \leqslant \|T^*\|$, and so we have shown that the equality $\|T\| = \|T^*\|$ must hold.

Of course, this last result is in agreement with the result previously established for an operator defined on any normed linear space and its adjoint operator (defined on the dual of the given normed linear space). However, it is of interest to note that the argument presented in the preceding paragraph enables us to avoid the use of the Hahn-Banach theorem.

Returning momentarily to Theorem 1, we see that (*b*) must be modified, in the present context, as follows: $(\alpha T)^* = \bar{\alpha} T$; parts (*a*) and (*d*) remain correct as stated, while part (*c*) merely assumes the form $I^* = I$.

We conclude this section with the following easy but important result.

Theorem 3: *If T is any operator on a Hilbert space H, then* $\|T^*T\| = \|TT^*\| = \|T\|^2$.

PROOF: For any unit vector f we have $\|T^*Tf\| \leqslant \|T^*\| \cdot \|Tf\| \leqslant \|T^*\| \cdot \|T\| \cdot \|f\| = \|T\|^2 \cdot \|f\| = \|T\|^2$; hence $\|T^*T\| \leqslant \|T\|^2$. On the other hand, for any unit vector f we also have $\|T^*Tf\| = \|T^*Tf\| \cdot \|f\| \geqslant |(T^*Tf, f)| = |(Tf, Tf)| = \|Tf\|^2$. Taking the supremum of the end terms over all unit vectors, we obtain $\|T^*T\| \geqslant \|T\|^2$. Thus, $\|T^*T\| = \|T\|^2$. Replacing T by T^*, we obtain $\|TT^*\| = \|T^*\|^2 = \|T\|^2$. This completes the proof.

Exercises

1. Prove that $(T^*)^*(f) = Tf$ for every vector f in N.
2. Prove directly (without reference to dual spaces) that $(T^*)^* = T$ for any operator T on a Hilbert space.

§3. THE INVERSE OF AN OPERATOR

Let A be any non-empty set whatsoever and let T be a mapping of A into A. If $T(A) = A$ (i.e., if the mapping T is *onto*) and if $T(x) \neq T(y)$ whenever $x \neq y$, then the inverse mapping T^{-1} is defined. If T^{-1} is temporarily denoted S, we readily see that $ST = I$ and $TS = I$, where I, of course, denotes the identity mapping on A. Conversely, it is not difficult to prove (cf. Exercise 1) that if S and T are mappings of A into A which satisfy *both* of the preceding equalities, then each of them is a one-to-one mapping of A onto A and that each of them is the inverse of the other. One might expect that either of the preceding equalities implies the other, but this is not so. For example, let A be the set of positive integers, let S be the mapping which sends every member n of A into $n + 1$, and let T send every integer n exceeding 1 into $n - 1$, while $T(1) = 7$. Then it is easily seen that $TS = I$ and $ST \neq I$. In fact, the mapping T is not even one-to-one, since $T(1) = T(8)$; on the other hand, S, while one-to-one, is not onto, since there exists no member n of A such that $S(n) = 1$.

If $ST = I$ we say that S is a left inverse of T and T is a right inverse of S. Thus, T^{-1} exists and equals S iff S is both a right and a left inverse of T. Now suppose that T has a left inverse S and a right inverse $\tilde{S}$; are S and $\tilde{S}$ necessarily the same mapping? The answer is in the affirmative, for we can argue as follows: $\tilde{S} = I\tilde{S} = (ST)\tilde{S} = S(T\tilde{S}) = SI = S$. (Note that we exploit the fact that composition of mappings is associative.) Thus, a mapping which possesses a left inverse and a right inverse possesses an inverse, and all three coincide.

Now let us turn to the case that the set A is a linear space V (not necessarily normed), and let T be a linear transformation on V. In order for T to be invertible it must, as remarked previously, be onto and one-to-one, but the linearity of T enables us to restate the second condition as follows: $Tf \neq o$ whenever $f \neq o$ (for, if $g \neq h$, then $T(g) - T(h) = T(g - h) \neq o$, and so $T(g) \neq T(h)$). Thus a linear transformation T on V is *invertible* iff $T(V) = V$ and $Tf = o$ holds *only* when $f = o$. It is then a triviality to show that the mapping T^{-1} is also linear.

However, if V is a *normed* linear space, N, and if T is a *bounded* linear transformation which is invertible, we may ask whether T^{-1} is necessarily bounded. The answer is in the negative, as is shown by the following example: let N consist of all sequences of scalars containing only a finite number of non-zero entries, let $\|(\alpha_1, \alpha_2, \alpha_3, \ldots)\| = \max_{1 \leqslant n < +\infty} |\alpha_n|$, and let $T(\alpha_1, \alpha_2, \alpha_3, \ldots) = (\alpha_1, \alpha_2/2, \alpha_3/3, \ldots)$. It is obvious that T is linear, one-to-one, onto, and bounded; in fact, $\|T\| = 1$. However, let a be the vector containing 1 in the k-th place and zeros everywhere else. Then it is immediately evident that $\|a\| = 1$, $T^{-1}a = ka$, $\|T^{-1}a\| = k\,\|a\| = k$, and so $\|T^{-1}\| \geqslant k$. Since k can be chosen as large as we wish, $\|T^{-1}\|$ is not finite—i.e., T^{-1} is a linear transformation but not an operator.

This unpleasant state of affairs disappears if instead of considering a normed linear space N we consider a Banach space B. (Note that the space considered in the preceding paragraph is not complete.) We shall first prove the following remarkable theorem and then obtain an easy corollary which, in turn, leads directly to the desired result.

Theorem 1: *Let T be an operator (not necessarily one-to-one) mapping the Banach space B onto B. Then for any positive number ϵ, $T(S_\epsilon(o))$ contains a neighborhood of o.*

PROOF: (*a*) For each positive integer n let $\tilde{S}_n$ denote $T(S_n(o))$. By hypothesis, $\bigcup_{n=1}^{\infty} \tilde{S}_n = B$. Since B is complete, the Baire category theorem guarantees that at least one $\tilde{S}_n$, say $\tilde{S}_k$, is not nowhere dense. Therefore, there exists a vector a and a positive number δ such that $\tilde{S}_k$ is dense in $S_\delta(a)$. Although a may fail to belong to $\tilde{S}_k$, it does belong to some $\tilde{S}_n$, say $a \in \tilde{S}_m$—i.e., $a = Tb$, where $\|b\| < m$.

(*b*) Now we shall show that $\tilde{S}_{k+m}$ is dense in some neighborhood of o. By the triangle inequality, $\|f\| < k$ implies that $\|f - b\| < k + m$, and so $\tilde{S}_{k+m}$ contains the set of all vectors expressible as $T(f - b)$ for some vector f of norm $<k$. By linearity, this set is precisely the set obtained by translating $\tilde{S}_k$ by $-a$, and this latter set is certainly dense in $S_\delta(o)$. Thus, $\tilde{S}_{k+m}$ is dense in some neighborhood of o—in fact, $\tilde{S}_{k+m}$ is dense in $S_\delta(o)$.

(*c*) By the homogeneity property of a linear transformation, $\tilde{S}_1$ is dense in $S_{\tilde{\delta}}(o)$, where $\tilde{\delta} = \delta/(k + m)$.

(*d*) Now we shall show that $\tilde{S}_2 \supseteq S_{\tilde{\delta}}(o)$; i.e., for any vector y in $S_{\tilde{\delta}}(o)$ we shall prove that there exists a vector g in $S_2(o)$ such that $Tg = y$. Since $\tilde{S}_1$ is dense in $S_{\tilde{\delta}}(o)$, we can find a vector y_1 in $\tilde{S}_1$ such that $\|y_1 - y\| < \tilde{\delta}/2$ and a vector x_1 in $S_1(o)$ such that $Tx_1 = y_1$. Thus, $\|y - Tx_1\| < \tilde{\delta}/2$, or $\|2y - 2Tx_1\| < \tilde{\delta}$. Repeating this argument, we see that we can find a vector x_2 in $S_1(o)$ such that $\|2y - 2Tx_1 - Tx_2\| < \tilde{\delta}/2$, or

$$\|4y - 4Tx_1 - 2Tx_2\| < \tilde{\delta}.$$

Similarly, we can find a vector x_3 in $S_1(o)$ such that

$$\|8y - 8Tx_1 - 4Tx_2 - 2Tx_3\| < \tilde{\delta},$$

and by induction we see that we can find vectors $x_4, x_5, \ldots$ in $S_1(o)$ such that $\|2^n y - 2^n Tx_1 - 2^{n-1}Tx_2 - \cdots - 2^1 Tx_n\| < \tilde{\delta}$, or $\|y - Tg_n\| < \tilde{\delta}/2^n$, where

$$g_n = x_1 + \frac{x_2}{2} + \frac{x_3}{2^2} + \cdots + \frac{x_n}{2^{n-1}}.$$

For $m > n$, we have

$$g_m - g_n = \frac{x_{n+1}}{2^n} + \cdots + \frac{x_m}{2^{m-1}},$$

and so

$$\|g_m - g_n\| \leqslant \frac{\|x_{n+1}\|}{2^n} + \cdots + \frac{\|x_m\|}{2^{m-1}} < \frac{1}{2^n} + \cdots + \frac{1}{2^{m-1}} < \sum_{k=n}^{\infty} \frac{1}{2^k} = \frac{1}{2^{n-1}}.$$

Hence, the g's form a Cauchy sequence, and so there exists a vector g such that $\|g - g_n\| \to 0$. Then, $\|y - Tg_n\| = \|(y - Tg) + T(g - g_n)\| < \tilde{\delta}/2^n$, and so $\|y - Tg\| \leqslant \|T(g - g_n)\| + \|(y - Tg) + T(g - g_n)\| \leqslant \|T(g - g_n)\| + \tilde{\delta}/2^n \leqslant \|T\| \cdot \|g - g_n\| + \tilde{\delta}/2^n$. Letting n increase without bound, we obtain $\|y - Tg\| = 0$, or $y = Tg$. Now,

$$\|g\| = \lim_{n\to\infty} \|g_n\| \leqslant \|x_1\| + \frac{\|x_2\|}{2} + \frac{\|x_3\|}{2^2} + \cdots < 1 + \tfrac{1}{2} + \frac{1}{2^2} + \cdots = 2.$$

Hence, $g \in S_2(o)$, and so $y \in \tilde{S}_2$. Thus, we have established the inclusion $\tilde{S}_2 \supseteq S_{\tilde{\delta}}(o)$. (Note that the boundedness of T is used only once, in claiming that $\|T(g - g_n)\| \to 0$, but this one point in the argument is vital.)

(*e*) Now, by homogeneity, $\tilde{S}_1 \supseteq S_{\tilde{\delta}/2}(o)$, and, again by homogeneity, the image of $S_\epsilon(o)$ contains $S_{\epsilon\tilde{\delta}/2}(o)$, which is a neighborhood of o. The proof is thus complete.

Corollary (Open Mapping Theorem): *If T satisfies the hypotheses of Theorem 1, it is an open mapping—that is, the image of an open set is open.*

PROOF: Let G be any open set in B, let $\tilde{G} = T(G)$, and let $\tilde{g} \in \tilde{G}$. Then there exists a vector g in G such that $Tg = \tilde{g}$. Since G is open, $\tilde{S}_\epsilon(g) \subseteq G$ for sufficiently small ϵ. Hence, for any vector h of norm $<\epsilon$, $T(g + h) \in \tilde{G}$, or $Tg + Th \in \tilde{G}$, or $\tilde{g} + Th \in \tilde{G}$. By Theorem 1, as h varies over $S_\epsilon(o)$, Th describes a set of vectors containing a neighborhood of o, and so $\tilde{g} + Th$ describes a set containing a neighborhood of $\tilde{g}$. Thus, $\tilde{g}$ has a neighborhood which is contained in $\tilde{G}$—in other words, g is an inner point of $\tilde{G}$. Thus $\tilde{G}$ is open, and so T is an open mapping.

We now obtain our objective with the following corollary.

Corollary: *Let the operator T furnish a one-to-one mapping of the Banach space B onto B. Then the inverse mapping T^{-1} is also an operator.*

PROOF: We have already noted that T^{-1} is linear. Since T is bounded, it is continuous, and so the preceding corollary guarantees that T is an open mapping. From §1–9, an open, continuous, one-to-one mapping of a metric space onto itself (or onto another metric space) possesses a continuous inverse. Hence T^{-1} is continuous; but for linear transformations continuity and boundedness are equivalent, and so T^{-1} is an operator.

As the final item in this section, we state a modified form of Theorem 2–1 of Chapter 5; we dispense with the proof, since it is virtually identical with the one given for the aforementioned theorem.

Theorem 2 (Banach-Steinhaus): *Let $\{T_\alpha\}$ be a collection of operators on a Banach space B, and suppose that for each vector f the numbers $\{\|T_\alpha f\|\}$ are bounded. Then the numbers $\{\|T_\alpha\|\}$ are bounded.*

Exercise

1. Prove the assertion made in the fourth sentence of the first paragraph of this section.

§4. SEQUENCES OF OPERATORS

Definition 1–5 and Theorem 1–5 show that the collection of all operators on a normed linear space N constitute a normed linear space, which we denote $\mathscr{B}(N)$. (The letter $\mathscr{B}$ serves to emphasize that we are dealing only with *bounded* linear transformations.) It is, therefore, possible to introduce into $\mathscr{B}(N)$ the concept of convergence. In fact, three different definitions of convergence arise quite naturally. The most obvious definition is that which comes directly from the definition of convergence in any metric space. We shall say that the sequence $\{T_1, T_2, T_3, \ldots\}$ converges *uniformly*, or in norm, to the operator T if $\|T - T_n\| \to 0$, and we shall employ the notation $T_n \Rightarrow T$, rather than $T_n \to T$, as might be expected.

Theorem 1: *If N is complete, so is* $\mathscr{B}(N)$.

PROOF: Let $\{T_1, T_2, T_3, \ldots\}$ be any Cauchy sequence in $\mathscr{B}(N)$. Then for any vector f in N the sequence $\{T_1f, T_2f, T_3f, \ldots\}$ is also Cauchy, for $\|T_nf - T_mf\| \leqslant \|T_n - T_m\| \cdot \|f\| \to 0$. Since N is complete, there exists a vector, which we denote Tf, such that $\|Tf - Tf_n\| \to 0$. Clearly $T(\alpha f + \beta g) = \alpha Tf + \beta Tg$ for all α, β, f, and g, so that T is a linear transformation. Furthermore, $\|Tf\| = \lim_{n\to\infty} \|T_nf\| \leqslant (\lim_{n\to\infty} \|T_n\|) \cdot \|f\|$, and so T is bounded, and hence a member of $\mathscr{B}(N)$. To show that $T_n \Rightarrow T$ we argue as follows. Given $\epsilon > 0$, there exists an index $M(\epsilon)$ such that $\|T_m - T_n\| < \epsilon$ whenever m and n both exceed $M(\epsilon)$. For any such m and n and for any vector f we may write $\|(T - T_m)f\| = \|(T - T_n)f + (T_n - T_m)f\| \leqslant \|(T - T_n)f\| + \|T_n - T_m\| \cdot \|f\| \leqslant \|(T - T_n)f\| + \epsilon \|f\|$. Letting n increase without bound while holding m fixed, we obtain $\|(T - T_m)f\| \leqslant \epsilon \|f\|$, and hence $\|T - T_m\| \leqslant \epsilon$ whenever $m > M(\epsilon)$. Therefore, $\mathscr{B}(N)$ is indeed complete.

Turning again to a sequence $\{T_1, T_2, T_3, \ldots\}$ of operators on the normed linear space N we note that it may happen that the sequence of vectors $\{T_1f, T_2f, T_3f, \ldots\}$ converges for every vector f and yet there may fail to exist an operator T such that $T_n \Rightarrow T$. We shall illustrate this possibility in the case of a complete normed linear space and then turn to the consideration of an additional complication which can arise when N is not complete.

Let us consider the Hilbert space l^2, and for any vector f, consisting of the sequence $(\alpha_1, \alpha_2, \alpha_3, \alpha_4, \ldots)$ of scalars, let $T_1f = (0, \alpha_2, \alpha_3, \alpha_4, \ldots)$,

$T_2f = (0, 0, \alpha_3, \alpha_4, \ldots)$, $T_3f = (0, 0, 0, \alpha_4, \ldots)$, etc. Clearly, $\|T_nf\| \to 0$, and so $T_nf \to o = Of$. Thus, T_n converges to the operator O in the sense that $\|(T_n - O)f\| \to 0$ for every vector f, but it is not true that $T_n \Rightarrow O$, for $\|T_n - O\| = \|T_n\|$, and each operator T_n obviously has unit norm. Thus, we are led to the following definition: The sequence of operators $T_1, T_2, T_3, \ldots$ on any normed linear space is said to *converge strongly*, or *pointwise*, if for every vector f the limit $\lim_{n\to\infty} T_nf$ exists. Since, obviously, the equality $\lim_{n\to\infty} T_n(\alpha f + \beta g) = \alpha \lim_{n\to\infty} T_nf + \beta \lim_{n\to\infty} T_ng$ holds for any choice of α, β, f, and g, the strongly convergent sequence T_n determines a linear transformation T such that $T_nf \to Tf$ for every vector f, and we write $T_n \to T$.

However, the question arises whether T is necessarily bounded. Theorem 3–2 immediately provides an affirmative answer if N is complete. On the other hand, the following simple example shows that T may be unbounded if N is not complete. Let N consist of those sequences of scalars in which only a finite number of the scalar components differ from zero, let the norm be defined as in l^2, and let the vectors T_nf be defined as follows for any vector $f = (\alpha_1, \alpha_2, \alpha_3, \alpha_4, \ldots)$:

$$T_1f = (\alpha_1, 2\alpha_2, \alpha_3, \alpha_4, \ldots),$$

$$T_2f = (\alpha_1, 2\alpha_2, 3\alpha_3, \alpha_4, \ldots),$$

$$T_3f = (\alpha_1, 2\alpha_2, 3\alpha_3, 4\alpha_4, \ldots), \text{ and so forth.}$$

Clearly, $T_nf \to (\beta_1, \beta_2, \beta_3, \beta_4, \ldots)$, where $\beta_k = k\alpha_k$ for all indices k. The linear transformation T thus defined on N is clearly unbounded.

We now turn to the third type of convergence, and we begin with a simple illustration. Once again let N be taken as l^2. Let $T_1f = (0, \alpha_1, \alpha_2, \alpha_3, \ldots)$, $T_2f = (0, 0, \alpha_1, \alpha_2, \alpha_3, \ldots)$, $T_3f = (0, 0, 0, \alpha_1, \alpha_2, \alpha_3, \ldots)$, and so forth, where, as before, $f = (\alpha_1, \alpha_2, \alpha_3, \ldots)$. It is easily seen that the sequence $\{T_n\}$ does not converge in either of the two senses defined previously, for if f is taken as the vector $(1, 0, 0, 0, \ldots)$ it is obvious that $\|T_nf - T_mf\| = 2^{1/2}$ whenever $n \neq m$, and so the sequence $\{T_nf\}$ does not converge; hence the sequence $\{T_n\}$ does not converge strongly, and so it certainly does not converge uniformly. Nevertheless, a certain type of convergence is involved here. Let l denote any bounded linear functional on l^2. Then we know that there exists a unique vector g, given by a sequence $(\gamma_1, \gamma_2, \gamma_3, \ldots)$, such that $l(f) = (f, g)$ and hence $l(T_nf) = (T_nf, g) = \sum_{k=1}^{\infty} \alpha_k\bar{\gamma}_{n+k}$. Employing the Schwarz inequality we obtain $|l(T_nf)|^2 \leqslant \{\sum_{k=1}^{\infty} |\alpha_k|^2\}\{\sum_{k=n+1}^{\infty} |\gamma_k|^2\}$. Letting n increase without bound, we conclude that $l(T_nf) \to 0 = l(Of)$.

We are thus led to the following definition: The sequence of operators $\{T_n\}$ is said to *converge weakly* to the operator T if for every vector f in N and every member l of N^* the limiting relation $l(T_nf) \to l(Tf)$ holds, and we write $T_n \rightharpoonup T$. Having defined three kinds of convergence of *sequences* of operators, we can analogously define three kinds of convergence of *series* of operators; we shall say that the series $S_1 + S_2 + S_3 + \cdots$ of operators converges to the operator S (uniformly, strongly, weakly) if the sequence $S_1, S_1 + S_2, S_1 + S_2 + S_3, \ldots$ of partial sums converges to S (uniformly, strongly, weakly). In particular, we shall employ several times the following simple but important theorem, whose proof is left as Exercise 1.

Theorem 2: *If the numerical series $\|S_1\| + \|S_2\| + \cdots$ is convergent, where the S_k's are operators on a complete normed linear space, then the series $S_1 + S_2 + S_3 + \cdots$ converges uniformly to an operator S. The sum S of the series is not affected by rearrangement of the terms of the series.*

Exercises

1. Prove Theorem 2.

2. Consider the following theorem: If the sequence of operators $\{S_n\}$ converges to the operator S and the sequence of operators $\{T_n\}$ converges to the operator T, then the sequence of operators $\{S_nT_n\}$ converges to the operator ST. This is not well formulated, since the kind of convergence is not specified. First, show that the theorem is true if the word "uniformly" is inserted after each appearance of the word "converges." Then determine, in as many cases as you can, the truth or falsity of the other 26 possible interpretations of this "theorem."

3. Prove the truth or falsity of the following theorems, where the operators T_n and T are defined on a normed linear space:

(*a*) If $T_n \Rightarrow T$, then $T_n^* \Rightarrow T^*$,
(*b*) If $T_n \rightarrow T$, then $T_n^* \rightarrow T^*$,
(*c*) If $T_n \rightharpoonup T$, then $T_n^* \rightharpoonup T^*$.

§5. HERMITIAN OPERATORS

In this section it is especially important to note that we are dealing with complex scalars; in particular, Theorem 1 and its corollary are false in the case of real Hilbert spaces. (Cf. Exercise 1.) A fixed Hilbert space H is under consideration.

Theorem 1: *Let T be an operator on H. If $(Tf, f) = 0$ for every vector f in H, then $T = O$.*

PROOF: Take any two vectors f, g. By hypothesis, $0 = (T(f + g), (f + g)) = (Tf, f) + (Tg, g) + (Tf, g) + (Tg, f) = 0 + 0 + (Tf, g) + (Tg, f)$. Thus, $(Tf, g) + (Tg, f) = 0$; replacing f by if, we obtain

$(Tf, g) - (Tg, f) = 0$. Adding these last two equations, we obtain $(Tf, g) = 0$; choosing f at pleasure and then choosing g as Tf, we obtain $(Tf, Tf) = 0$, or $Tf = o$. Thus, $T = O$.

Corollary: *If $(T_1 f, f) = (T_2 f, f)$ for all vectors f, then $T_1 = T_2$.*

PROOF: Rewrite the preceding equality in the form

$$((T_1 - T_2)f, f) = 0$$

and apply the preceding theorem.

Definition 1: *An operator T is said to be hermitian, or self-adjoint, if $T = T^*$. (Trivial examples are provided by the operators O and I; later we shall encounter less obvious examples.)*

Theorem 2: *(a) The sum of any finite number of hermitian operators is hermitian.*

(b) The limit (weak, strong, or uniform) of hermitian operators is hermitian.

(c) A real scalar multiple of a hermitian operator is hermitian.

(d) The product of two hermitian operators is hermitian iff the given operators commute.

PROOF: Left as Exercise 2.

Theorem 3: *The operator T is hermitian iff (Tf, f) is real for all vectors f.*

PROOF: (*a*) If T is hermitian, then, for every vector f, $(Tf, f) = (f, Tf) = \overline{(Tf, f)}$. Hence (Tf, f) equals its conjugate, and so it is real.

(*b*) If (Tf, f) is always real, then $(Tf, f) = \overline{(Tf, f)} = \overline{(f, T^*f)} = (T^*f, f)$. By the corollary to Theorem 1, $T = T^*$, and so T is hermitian.

Theorem 4: *If T is hermitian, so are all the operators $T^2, T^3, \ldots$.*

PROOF: Since T commutes with itself, part (*d*) of Theorem 2 guarantees that T^2 is hermitian; since T and T^2 commute, this same result guarantees that T^3 is hermitian. Since this argument can be repeated indefinitely, we obtain the desired result.

Theorem 5: *If T is hermitian and $\neq O$, then no power of T is O. (That is, a hermitian operator cannot be nilpotent.)*

PROOF: By hypothesis there exists a vector f such that $Tf \neq o$. Hence $0 < (Tf, Tf) = (T^2f, f)$, and so $T^2f \neq o$. Similarly, $T^4f = T^2(T^2f) \neq o$, $T^8f \neq o$, and so forth. If $T^m f = o$ for *some* positive integer m, we could choose the integer k so large that $2^k > m$, and we would obtain $T^{2^k}f = T^{2^k-m}(T^m f) = T^{2^k-m}o = o$, contradicting the result obtained in the previous sentence.

An alternative proof goes as follows: For any operator T, hermitian or not, we have seen that $\|T^*T\| = \|T\|^2$. Thus, if T is hermitian, we obtain $\|T^2\| = \|T\|^2$, and so $T^2 \neq O$; similarly, $T^4 \neq O$, $T^8 \neq O$, and so forth. (Cf. Exercise 3.)

Theorem 6: *If T is hermitian,* $\|T\| = \sup_{\|f\|=1} |(Tf, f)|$.

PROOF: (*a*) Denote the right side of the equality by μ. Then for any unit vector f, $|(Tf, f)| \leqslant \|Tf\| \cdot \|f\| \leqslant \|T\| \cdot \|f\| \cdot \|f\| = \|T\|$. Letting f vary freely under the sole restriction $\|f\| = 1$, we obtain $\mu \leqslant \|T\|$. (Note that this half of the proof does not use the hermitian character of T.)

(*b*) Let f be any vector and let λ be any real positive number. Taking account of the definition of μ and of its obvious consequence $|(Tg, g)| \leqslant \mu \|g\|^2$, we obtain

$$|(T(\lambda f \pm \lambda^{-1}Tf, \lambda f \pm \lambda^{-1}Tf)| \leqslant \mu \|\lambda f \pm \lambda^{-1}Tf\|^2$$

and so (using the triangle inequality for real numbers)

$$|(T(\lambda f + \lambda^{-1}Tf, \lambda f + \lambda^{-1}Tf) - (T(\lambda f - \lambda^{-1}Tf), \lambda f - \lambda^{-1}Tf)|$$
$$\leqslant \mu\{\|\lambda f + \lambda^{-1}Tf\|^2 + \|\lambda f - \lambda^{-1}Tf\|^2\}.$$

A routine calculation shows that the left side is simply $4\|Tf\|^2$. (In this calculation the hermitian character of T is used.) The right side reduces, by the parallelogram law, to $2\mu(\lambda^2 \|f\|^2 + \lambda^{-2}\|Tf\|^2)$. Thus, for any *unit* vector f and any positive number λ we obtain

$$4\|Tf\|^2 \leqslant 2\mu(\lambda^2 + \lambda^{-2}\|Tf\|^2).$$

A trivial calculation shows that the right side, considered as a function of λ, is minimized by choosing $\lambda = \|Tf\|^{1/2}$. Since the left side of the inequality is independent of λ, we obtain $4\|Tf\|^2 \leqslant 4\mu\|Tf\|$. Dividing by $4\|Tf\|$ we obtain $\|Tf\| \leqslant \mu$, and so $\|T\| \leqslant \mu$; combining this result with that obtained in part (*a*), we conclude that $\|T\| = \mu$. (The division by $4\|Tf\|$ is not legitimate if $\|Tf\| = 0$, but in this case the desired inequality, namely $\|Tf\| \leqslant \mu$, is trivially true.)

Corollary: *If T is hermitian,* $\|T\| = \max\{|m|, |M|\}$, *where*

$$m = \inf_{\|f\|=1} (Tf, f), \quad M = \sup_{\|f\|=1} (Tf, f).$$

PROOF: Obvious from Theorem 6.

Definition 2: *A hermitian operator T is said to be positive if $(Tf, f) \geqslant 0$ for all vectors f. (Examples: $T = O$, $T = I$.) Obviously, $\|T\| = M$ in this case, where M is defined in the preceding corollary.*

This definition suggests the possibility of introducing an ordering into the family of hermitian operators, analogous to the relation $\leqslant$ which exists in R. Actually, only a *partial* ordering is provided by the following definition. (Cf. Exercise 4.)

Definition 3: *If T_1 and T_2 are hermitian, then we write $T_1 \geqslant T_2$ iff $T_1 - T_2$ is positive. (In particular, $T \geqslant O$ iff T is positive.)*

Theorem 7: *If $T_1 \geqslant T_2$ and $T_3 \geqslant T_4$ and if α and β are non-negative real numbers, then $\alpha T_1 + \beta T_3 \geqslant \alpha T_2 + \beta T_4$. Also, $mI \leqslant T \leqslant MI$, where m and M are the numbers defined in the corollary to Theorem* 6. (*Cf. Exercise* 5.)

PROOF: Trivial.

It should be noted that the product of positive operators need not be positive, for the operators may fail to commute, in which case their product is not hermitian. However, it is an interesting (and non-trivial) fact that the product of *commuting* positive operators is indeed positive. This will be proven in Theorem 9; first we need the following interesting theorem, however.

Theorem 8: *If $T \geqslant O$, there exists a positive hermitian operator S satisfying $S^2 = T$. (That is, every positive operator has at least one positive square root.)*

PROOF: Trivial if $T = O$. If $T \neq O$, we may confine attention to the case that $\|T\| = 1$ (for otherwise we can work with $(1/\|T\|)T$, whose norm is 1, and multiply the square root obtained in this case by $\|T\|^{1/2}$). Then, clearly, $O \leqslant T \leqslant I$ and $O \leqslant I - T \leqslant I$.

Now we recall from calculus the Taylor expansion

$$(1 - u)^{1/2} = 1 - c_1 u - c_2 u^2 - \cdots,$$

where the c's are *positive*. (Their exact values can be written down, but are of no importance.) This expansion is easily shown to be valid for $-1 < u < 1$, but we shall show that it is, in fact, also valid for $u = 1$. (Also for $u = -1$, but this is of no importance here.) Since the c's are all positive, we can speak of their sum, if we allow $+\infty$. Now, if the sum were $+\infty$, or even any finite number >1, a *finite* number of the c's would add up to more than 1, say $c_1 + c_2 + \cdots + c_n > 1$. Then, by *continuity*, $c_1 u + c_2 u^2 + \cdots + c_n u^n > 1$ for u sufficiently close to 1, and so we would have $c_1 u + c_2 u^2 + \cdots + c_n u^n + \cdots > 1$. But this would say that $1 - c_1 u - c_2 u^2 - \cdots$ is *negative* when u is sufficiently close to 1, and so we obtain the absurd inequality $(1 - u)^{1/2} < 0$ for u close to 1. Thus, the series $c_1 + c_2 + \cdots + c_n + \cdots$ adds up to 1 *or less*. But if the sum were *less* than one, $c_1 u + c_2 u^2 + \cdots + c_n u^n + \cdots$ would be bounded away from 1 for $0 < u < 1$, and so $(1 - u)^{1/2}$ would *not* approach zero as $u \to 1$; this is also absurd. Hence, $c_1 + c_2 + \cdots = 1$, and so the series expansion given for $(1 - u)^{1/2}$ is correct for $u = 1$. Now, let us write out the series

$$I - c_1(I - T) - c_2(I - T)^2 - \cdots.$$

The partial sums $(T_n = I - c_1(I - T) - c_2(I - T)^2 - \cdots - c_n(I - T)^n)$ are a sequence of hermitian (why?) operators. Furthermore, if $m > n$,

$$\|T_m - T_n\| = \|c_{n+1}(I - T)^{n+1} + c_{n+2}(I - T)^{n+2} + \cdots + c_m(I - T)^m\|$$

$$\leqslant \sum_{k=n+1}^{m} \|c_k(I - T)^{n+1}\| = \sum_{k=n+1}^{m} c_k \|I - T\|^{n+1} \leqslant \sum_{k=n+1}^{m} c_k.$$

Since $\sum_{k=1}^{\infty} c_k$ is convergent, the sum $\sum_{k=n+1}^{m} c_k$ can be made $<\epsilon$ by choosing $n > N(\epsilon)$. Hence, the sequence of partial sums $T_1, T_2, \ldots$ converges (in norm) to a hermitian operator S. Thus, we may write

$$S = I - c_1(I - T) - c_2(I - T)^2 - \cdots.$$

Now, if we square the series expansion $1 - c_1 u - c_2 u^2 - \cdots$ we have $1 - u$, and similarly we find that $S^2 = I - (I - T) = T$. Furthermore, since $c_1 + c_2 + \cdots + c_n + \cdots = 1$, it immediately follows that

$$\|c_1(I - T) + c_2(I - T)^2 + \cdots\| \leqslant 1,$$

and so $O \leqslant c_1(I - T) + c_2(I - T)^2 + \cdots \leqslant I$. Therefore, $O \leqslant I - c_1(I - T) - c_2(I - T)^2 - \cdots = S \leqslant I$, and so S is a positive operator.

Theorem 9: *If A and B are positive commuting operators, then AB is positive.*

PROOF: The hermitian character of AB is assured by Theorem 2. We construct positive square roots S_1 and S_2 of A and B, respectively, in the manner described in the preceding proof. Since B commutes with A, B commutes with all polynomials in A; this suggests that B commutes with S_1. We demonstrate this as follows: Let $Q_1, Q_2, Q_3, \ldots$ be the partial sums of the infinite series representing S_1. Then $BS_1 - S_1B = BQ_n + B(S_1 - Q_n) - (S_1 - Q_n)B - Q_nB = B(S_1 - Q_n) - (S_1 - Q_n)B$. Hence,

$$\|BS_1 - S_1B\| \leqslant \|B(S_1 - Q_n)\| + \|(S_1 - Q_n)B\|$$
$$\leqslant \|B\| \cdot \|S_1 - Q_n\| + \|S_1 - Q_n\| \cdot \|B\| = 2\,\|B\| \cdot \|S_1 - Q_n\| \to 0,$$

and so $BS_1 - S_1B = O$, or $BS_1 = S_1B$.

Similarly, it follows that S_1 commutes with S_2. Therefore, S_1S_2 is hermitian; furthermore, $(S_1S_2)^2 = S_1S_2S_1S_2 = S_1S_1S_2S_2 = S_1^2S_2^2 = AB$. For any vector f, $(ABf, f) = ((S_1S_2)^2f, f) = (S_1S_2f, S_1S_2f) \geqslant 0$, and so AB is positive.

Theorem 10: *A positive operator T possesses a unique positive square root.*

PROOF: Let S be the square root of T constructed in the proof of Theorem 8, and suppose that S_1 is *any* positive square root of T; we shall show that $S = S_1$. Since $S_1T = S_1(S_1^2) = (S_1^2)S_1 = TS_1$, we see that S_1 commutes with T. By the argument employed in the preceding proof, we see that S_1 commutes with S. Now we construct positive square roots of S and S_1 in the manner employed in the proof of Theorem 8; we call these square roots R and R_1, respectively. For any vector f, we obtain (using the fact that S_1 and S commute and also the fact that all operators involved are hermitian):

$$\begin{aligned}\|R(S - S_1)f\|^2 + \|R_1(S - S_1)f\|^2 &= (R^2(S - S_1)f, (S - S_1)f)\\ &\quad + (R_1^2(S - S_1)f, (S - S_1)f)\\ &= (S(S - S_1)f, (S - S_1)f)\\ &\quad + (S_1(S - S_1)f, (S - S_1)f)\\ &= ((S + S_1)(S - S_1)f, (S - S_1)f)\\ &= ((S^2 - S_1^2)f, (S - S_1)f)\\ &= ((T - T)f, (S - S_1)f)\\ &= (o, (S - S_1)f) = 0.\end{aligned}$$

Thus $R(S - S_1) = O = R_1(S - S_1)$, and so $R^2(S - S_1) = O = R_1^2(S - S_1)$;

this may be rewritten in the form $S(S - S_1) = S_1(S - S_1)$, or $(S - S_1)^2 = O$. Since, as we have seen, a hermitian operator cannot be nilpotent, it follows that $S - S_1 = O$, or $S = S_1$.

Having established this uniqueness result, we are now justified in referring to *the* positive square root of a given positive operator T, which we shall denote $T^{1/2}$. If we take *any* hermitian operator A, A^2 is positive, and so $(A^2)^{1/2}$ is the *unique* positive solution of the operator equality $X^2 = A^2$. For obvious reasons, we denote this operator as the absolute value of A: $|A| = (A^2)^{1/2}$.

We conclude this section by stating without proof the following theorem, which serves to emphasize the similarity between the hermitian operators and the real numbers when considered as members of $\mathscr{B}(H)$ and $\mathfrak{C}$, respectively.

Theorem 11: *Given any hermitian operator A, the operators $\frac{1}{2}(|A| \pm A)$ are positive, they commute, and their product is the zero operator. (For obvious reasons, we call $\frac{1}{2}(|A| + A)$ and $\frac{1}{2}(|A| - A)$ the positive part and negative part of A, respectively, and we denote them A^+ and A^-; thus, $A = A^+ - A^-$.)*

Exercises

1. Prove that Theorem 1 and its corollary are false in a *real* Hilbert space.
2. Prove Theorem 2.
3. Prove that the equality $\|T^m\| = \|T\|^m$ holds for any positive integer m if T is hermitian.
4. Show that two hermitian operators A, B may be incomparable—i.e., neither $A \geqslant B$ nor $B \geqslant A$ may be true.
5. Prove that if $T_1 \geqslant T_2$ and $T_2 \geqslant T_1$, then $T_1 = T_2$.
6. Prove the generalized Schwarz inequality: If T is a positive operator, then $|(Tf, g)|^2 \leqslant (Tf, f)(Tg, g)$ for all vectors f and g. (Thus, if $(Tf, f) = 0$, it follows that $(Tf, g) = 0$ for all vectors g; in particular, $(Tf, Tf) = 0$, or $Tf = o$. Therefore, a positive operator cannot map a vector f into a vector orthogonal to f except by annihilating f.)

§6. PROJECTIONS

Although the concept of a projection is of significance in Banach spaces, and even in linear spaces which are not provided with a norm, we shall confine attention entirely to Hilbert spaces.

Definition 1: *Let M be any subspace of H and let each vector f of H be expressed (in accordance with the projection theorem) in the form $f = g + h$, where $g \in M$ and $h \in M^{\perp}$. We call g the projection (more strictly, the orthogonal projection) of f on M, and we write $g = P_M f$. (Similarly, of course, $h = P_N f$, where $N = M^{\perp}$.)*

Theorem 1: *P_M is an operator, its norm is 1 except when $M = \{o\}$, and it is hermitian, positive, and idempotent (i.e., $P_M{}^2 = P_M$).*

PROOF: If $f = g + h$ is the break-up of f, then $\alpha f = (\alpha g) + (\alpha h)$ is a legitimate break-up of αf; by uniqueness, it is the *only* break-up of αf, and so $P_M(\alpha f) = \alpha P_M f$. Similarly, $P_M(f_1 + f_2) = P_M f_1 + P_M f_2$. Thus, P_M is linear. Since $\|f\|^2 = \|g\|^2 + \|h\|^2$, we have $\|g\| \leqslant \|f\|$, or $\|P_M f\| \leqslant \|f\|$. Thus, P_M is bounded; in fact, $\|P_M\| \leqslant 1$. If M contains a non-zero vector f, then for this particular f we have the break-up $f = f + o$, and so $\|P_M f\| = \|f\|$, showing that $\|P_M\| \geqslant 1$. Thus, $\|P_M\|$ is *exactly* 1 except when $M = \{o\}$, in which case, of course, $P_M = O$. Let $f_1 = g_1 + h_1$ and $f_2 = g_2 + h_2$. Then $(P_M f_1, f_2) - (f_1, P_M f_2) = (g_1, g_2 + h_2) - (g_1 + h_1, g_2) = (g_1, g_2) + (g_1, h_2) - (g_1, g_2) - (h_1, g_2) = (g_1, h_2) - (h_1, g_2) = 0$, since $(g_1, h_2) = 0$ and $(h_1, g_2) = 0$. Thus, for all vectors f_1 and f_2, $(P_M f_1, f_2) = (f_1, P_M f_2)$, and so $P_M^* = P_M$. Finally, for any f, we have the break-up $f = g + h$, where $g = P_M f$, and for g we obviously have the break-up $g = g + o$, so that $g = P_M g$. Thus, $P_M f = P_M(P_M f) = P_M{}^2 f$. Hence, $P_M = P_M{}^2$. Since $P_M{}^2$ is a positive operator, so is P_M (*because* $P_M = P_M{}^2$).

Theorem 2: *An operator P is a projection iff it is hermitian and idempotent.*

PROOF: We already have half the result—if P is a projection, it must be hermitian and idempotent. Now suppose P is hermitian and idempotent. Let M be the range of P. For every f in M, it is possible to find a vector g in H (perhaps more than one) such that $Pg = f$. But then $Pf = PPg = P^2 g = Pg = f$. Thus, every member of M is unchanged by the application of P. Furthermore, if $f = Pf$, then $f \in M$ (from the very definition of M). Thus, M consists *precisely* of those vectors which satisfy $f = Pf$. If f_1 and f_2 each belong to M, then $P(\alpha f_1 + \beta f_2) = \alpha P f_1 + \beta P f_2 = \alpha f_1 + \beta f_2$, and so $\alpha f_1 + \beta f_2$ is unchanged by P; thus M is a linear manifold. Finally, if $f_1, f_2, \ldots$ all belong to M and if this sequence converges to f, we have $Pf = P(f - f_n) + Pf_n = P(f - f_n) + f_n$; hence, $f_n - Pf = P(f_n - f)$, and so $\|f_n - Pf\| = \|P(f_n - f)\| \leqslant \|P\| \cdot \|f_n - f\| \to 0$. Therefore $\|f_n - Pf\| \to 0$, and so $f_n \to Pf$. But $f_n \to f$, and so $f = Pf$ (since limits are unique). Thus, M is closed, and so M is a subspace.

Now, let $h \in M^{\perp}$. Since P is hermitian, $(Ph, Ph) = (h, P^2 h)$, and since $P = P^2$, $(Ph, Ph) = (h, Ph)$. But $h \in M^{\perp}$ and $Ph \in M$, and so $(h, Ph) = 0$. Therefore, $(Ph, Ph) = 0$, and so $Ph = o$. To sum up,

$Pg = g$ for *every* vector g in M, and $Ph = o$ for *every* vector h in $M^\perp$. Then, for *any* vector f, we can write $f = g + h$, where $g \in M$ and $h \in M^\perp$ (by the projection theorem), and so $Pf = Pg + Ph = g + o = g$; that is, $Pf = P_M f$ (where P_M denotes projection upon M). Hence, $P = P_M$, and so P really is a projection—in fact, it is the projection upon its range.

Corollary: (*a*) *If P is a projection so is $I - P$, and conversely.*
(*b*) *If P is the projection upon M, then $I - P$ is the projection upon $M^\perp$.*

PROOF: If P is a projection, P is hermitian, and so $(I - P)^* = I^* - P^* = I - P$, so that $I - P$ is hermitian. Furthermore, $(I - P)^2 = I - 2P + P^2 = I - 2P + P = I - P$, and so $I - P$ is idempotent. By Theorem 2, $I - P$ is a projection. Conversely, if $I - P$ is a projection, so is $I - (I - P) = P$. This completes the proof of part (*a*). The proof of part (*b*) is left as Exercise 1.

Theorem 3: (*a*) *If P and Q are projections, then PQ is a projection iff $PQ = QP$.*
(*b*) *If P and Q are projections, then $P + Q$ is a projection iff $PQ = O$.*
(*c*) *If P and Q are projections, then $P - Q$ is a projection iff $PQ = Q$.*

PROOF: (*a*) If $PQ = QP$, then PQ is hermitian by Theorem 5–2; furthermore, $(PQ)^2 = PQPQ = PPQQ = P^2Q^2 = PQ$, and so PQ is idempotent. Theorem 2 now guarantees that PQ is a projection. If $PQ \neq QP$, then PQ is not hermitian (by Theorem 5–2), and so it is certainly not a projection.

(*b*) $P + Q$ is certainly hermitian, and $(P + Q)^2 = P^2 + QP + PQ + Q^2 = P + Q + QP + PQ$. Thus, $P + Q$ is a projection iff $QP + PQ = O$. If $PQ = O$, then $O = O^* = (PQ)^* = Q^*P^* = QP$, and so $PQ + QP = O$. Thus, the condition $PQ = O$ implies that $P + Q$ is a projection. On the other hand, if $PQ + QP = O$, multiplication on the left by P (and use of the equality $P^2 = P$) furnishes the equality $PQ + PQP = O$; similarly, multiplication on the right by P furnishes the equality $PQP + QP = O$. Subtraction of the last two equalities furnishes the result $PQ - QP = O$, or $PQ = QP$. Hence $O = PQ + QP = PQ + PQ = 2PQ$, or $PQ = O$. This completes the proof of part (*b*).

(*c*) By the corollary to Theorem 2, $P - Q$ is a projection iff $I - (P - Q)$, or $(I - P) + Q$, is a projection. Employing part (*b*) of the present theorem, with P replaced by $I - P$, we see that $P - Q$ is a projection iff $(I - P)Q = O$, or $Q = PQ$.

The three parts of Theorem 3 have all been established by "pushing around" symbols, but it is very important to visualize the geometrical significance of the various conditions that are imposed upon the operators P and Q. Careful attention to Exercise 2 is therefore strongly advised.

Exercises

1. Prove part (*b*) of the corollary to Theorem 2.

2. Let P and Q be projections associated with subspaces M_P and M_Q respectively. Keep in mind that M_P and M_Q are themselves Hilbert spaces.

(*a*) Prove that $PQ = QP$ iff the orthogonal complement of $M_P \cap M_Q$ with respect to M_P and the orthogonal complement of $M_P \cap M_Q$ with respect to M_Q are orthogonal to each other. Also, show that when this condition is satisfied, PQ is the projection associated with the subspace $M_P \cap M_Q$. (Hint: Try to visualize the situation when H is euclidean 3-space.)

(*b*) Prove that $PQ = O$ iff M_P and M_Q are orthogonal; equivalent formulations are that $M_P \subseteq (M_Q)^\perp$ and that $M_Q \subseteq (M_P)^\perp$. When this condition is satisfied, show that $P + Q$ is the projection on the *direct sum* of M_P and M_Q; by this expression we mean the set of all vectors expressible as the sum of a vector belonging to M_P and a vector belonging to M_Q. (Such an expression is unique, and the direct sum is indeed a subspace; prove these statements.)

(*c*) Prove that $PQ = Q$ iff $M_Q \subseteq M_P$, and that when this condition is satisfied $P - Q$ is the projection associated with the subspace $M_P \cap (M_Q)^\perp$ (i.e., the set of all vectors in M_P which are orthogonal to M_Q).

§7. THE SPECTRUM OF AN OPERATOR

We shall discuss in this section some of the basic ideas relating to the spectrum of an operator on a Banach space B. Recall that the operator T is invertible iff it is one-to-one and onto; the boundedness of T^{-1} need not be assumed, since it is guaranteed by the second corollary to Theorem 3–1.

Theorem 1: *If the operators T_1 and T_2 are both invertible, then so is T_1T_2, and $(T_1T_2)^{-1} = T_2^{-1}T_1^{-1}$.*

PROOF: Merely observe that $T_2^{-1}T_1^{-1}$ is both a left and right inverse of T_1T_2.

Theorem 2: *If the operators T_1 and T_2 commute and if T_1T_2 is invertible, then T_1* and *T_2 are both invertible.* (*Cf. Exercise* 1.)

PROOF: Let the inverse of T_1T_2, whose existence is assumed, be denoted S. Then $T_1T_2S = I$ and $ST_1T_2 = I$. The latter equation may be rewritten $ST_2T_1 = I$, since T_1 and T_2 are assumed to commute. Thus, T_1

possesses a left inverse (namely ST_2) and a right inverse (namely T_2S). Hence, $ST_2 = T_2S = T_1^{-1}$. Similarly $ST_1 = T_1S = T_2^{-1}$.

Theorem 3: *If $\|T - I\| < 1$, then T^{-1} exists; equivalently, if $\|S\| < 1$, then $(I - S)^{-1}$ exists.*

PROOF: We work with the second formulation of the theorem. The terms of the series $I + S + S^2 + S^3 + \cdots$ are dominated in norm respectively by the terms of the series $1 + \|S\| + \|S\|^2 + \|S\|^3 + \cdots$. The latter series converges (since $\|S\| < 1$) and so the original series converges to an operator which we term $\tilde{S}$. Then $\tilde{S}(I - S) = \tilde{S} - \tilde{S}S = (I + S + S^2 + S^3 + \cdots) - (S + S^2 + S^3 + \cdots) = I$, and so $\tilde{S}$ is a left inverse of $I - S$. Similarly, $\tilde{S}$ is a right inverse of $I - S$. Hence $(I - S)^{-1}$ exists and equals $\tilde{S}$.

Corollary: *If T is invertible there exists a positive number ϵ such that all operators $\tilde{T}$ satisfying the condition $\|\tilde{T} - T\| < \epsilon$ are also invertible; in fact, we may choose $\epsilon = 1/\|T^{-1}\|$.*

PROOF: Since T is invertible, $\tilde{T} = T - (T - \tilde{T}) = T[I - T^{-1}(T - \tilde{T})]$. If $\|T - \tilde{T}\| < 1/\|T^{-1}\|$, the inequalities $\|T^{-1}(T - \tilde{T})\| \leqslant \|T^{-1}\| \cdot \|T - \tilde{T}\| < 1$ hold, and so $I - T^{-1}(T - \tilde{T})$ is invertible. It then follows from Theorem 1 that $\tilde{T}$ is also invertible.

It should be pointed out that Theorem 3 is very closely related to the contraction theorem of Chapter 1. The equation $(I - S)f = g$ may be rewritten in the form $f = g + Sf$. The (non-linear) mapping $f \to g + Sf$ is contracting for any choice of the vector g, since $\|(g + Sf_1) - (g + Sf_2)\| = \|S(f_1 - f_2)\| \leqslant \|S\| \cdot \|f_1 - f_2\|$ and $\|S\| < 1$ by hypothesis. Since we are dealing with a complete metric space, the contraction theorem guarantees that the equation $f = g + Sf$ possesses a unique solution; thus $I - S$ is indeed invertible. If we use the method of iteration, employing the initial guess o, we obtain the following sequence of approximations: $o, g, g + Sg, g + Sg + S^2g, \ldots$, and so in the limit the solution $f = (I + S + S^2 + S^3 + \cdots)g$ is obtained, in agreement with the formula obtained for $(I - S)^{-1}$.

Definition 1: *The set of values of the scalar λ for which $(\lambda I - T)^{-1}$ exists is called the resolvent set of the operator T; the remaining values of λ constitute the spectrum of T, denoted $\sigma(T)$.*

Theorem 4: *If $|\lambda| > \|T\|$, the operator $\lambda I - T$ is invertible.*

PROOF: $\lambda I - T = (\lambda I)(I - \lambda^{-1}T)$. Since $\|\lambda^{-1}T\| = \|T\|/|\lambda| < 1$, Theorem 3 implies that $I - \lambda^{-1}T$ is invertible, and then Theorem 1 guarantees that $\lambda I - T$ is invertible. (Note carefully that $\lambda I - T$ *may* be invertible when $|\lambda| \leqslant \|T\|$; for example, if $T = I$, $\lambda I - T$ is invertible when $\lambda = \frac{1}{2}$, although $|\lambda| < \|T\| = 1$.)

Thus, the spectrum of T is confined to the closed disc $\{\lambda \mid |\lambda| \leqslant \|T\|\}$ of the complex plane. Since the corollary to Theorem 3 is obviously equivalent to the assertion that the resolvent set of T is open, we obtain the following result.

Theorem 5: *$\sigma(T)$ is a compact subset of the closed disc*

$$\{\lambda \mid |\lambda| \leqslant \|T\|\}.$$

It is conceivable, of course, that the spectrum may be empty, and, in fact, this can actually occur in *real* Banach spaces. (Cf. Exercise 2.) However, in Appendix B we show that the spectrum of an operator on any *complex* Banach space cannot be empty. The proof will be readily understood by the reader if he is acquainted with the elements of the theory of analytic functions (through Liouville's theorem).

Definition 2: *If there exists a non-zero vector f and a scalar μ such that $Tf = \mu f$, μ is called a characteristic value, or eigenvalue, of T, and f is called a characteristic vector, or eigenvector, of T. Since $Tf = \mu f$ and $Tf = \nu f$ obviously imply that $\mu = \nu$ (if $f \neq o$), we may refer to μ as the eigenvalue of T associated with f.*

We conclude this section with the observation that any eigenvalue of an operator T belongs to the spectrum of T, for if $Tf = \mu f$ ($f \neq o$), then $(\mu I - T)f = (\mu I - T)o = o$, and so $\mu I - T$ is not one-to-one. However, as shown in Exercise 3, the spectrum may contain numbers which are not eigenvalues.

Exercises

1. Prove that it is not permissible to drop the commutativity condition in Theorem 2.

2. Give an example of an operator on a real Banach space possessing empty spectrum.

3. Consider the (complex) Hilbert space $L^2([0, 1])$, and let $Tf = xf$. (More precisely, $Tf = g$, where $g(x) = xf(x)$ almost everywhere.) Show that (*a*) T is hermitian, (*b*) $\|T\| = 1$, (*c*) T possesses no eigenvalues, and (*d*) the spectrum of T consists precisely of the interval $[0, 1]$.

§8. SPECTRA OF HERMITIAN, NORMAL, AND UNITARY OPERATORS

In this section we present some elementary but important facts concerning the spectra of certain types of operators on a Hilbert space.

Theorem 1: *The eigenvalues of a hermitian operator T are real; if g_1 and g_2 are eigenvectors of T corresponding to distinct eigenvalues μ_1 and μ_2, respectively, then $(g_1, g_2) = 0$.*

PROOF: (*a*) Let μ be any eigenvector of T and let g be an eigenvector associated with μ. Then $(Tg, g) = (\mu g, g) = \mu(g, g)$, and so $\mu = (Tg, g)/(g, g)$. Since (Tg, g) is real, μ is the quotient of two real numbers, and hence is real.

(*b*) Since T is hermitian, $(Tg_1, g_2) = (g_1, Tg_2)$, and so $(\mu_1 g_1, g_2) = (g_1, \mu_2 g_2)$. Since μ_2 (like μ_1) is real, we obtain $\mu_1(g_1, g_2) = \mu_2(g_1, g_2)$, or $(\mu_1 - \mu_2)(g_1, g_2) = 0$. Since $\mu_1 \neq \mu_2$, it follows that $(g_1, g_2) = 0$.

Theorem 2: *The spectrum of a hermitian operator T is confined entirely to the real axis. (This result contains the first part of Theorem* 1.)

PROOF: (*a*) If μ is not real, then the preceding theorem shows that the equation $(\mu I - T)f = o$ implies that $f = o$. Therefore, $\mu I - T$ is certainly one-to-one.

(*b*) Next we shall prove that the range of $\mu I - T$ is a dense subset of H whenever μ is not real. For, in the contrary case, the closure of the range of $\mu I - T$ would be a *proper* subspace of H. By the projection theorem, there would exist a non-zero vector h orthogonal to S, and hence orthogonal to the range of $\mu I - T$. Thus, for every vector f,

$$(h, (\mu I - T)f) = 0;$$

choosing $f = h$, we obtain $(h, \mu h) = (h, Th)$, and so $\bar{\mu} = (h, Th)/(h, h)$. Since T is hermitian, $\bar{\mu}$, and hence μ, would be real, contrary to hypothesis.

(*c*) Now let f be any vector. From (*b*) we know that there exists a sequence of vectors $f_1, f_2, f_3, \ldots$ belonging to the range of $\mu I - T$ and converging to f. Therefore, we can find a sequence of vectors $g_1, g_2, g_3, \ldots$ such that $(\mu I - T)g_n = f_n$. The sequence $(\mu I - T)g_n$, being convergent, is Cauchy; therefore, given $\epsilon > 0$, there exists an index $N(\epsilon)$ such that $\epsilon > \|(\mu I - T)(g_n - g_m)\|$ whenever m and n both exceed $N(\epsilon)$. Now, a simple calculation (cf. Exercise 1) shows that

$$\|(\mu I - T)(g_n - g_m)\|^2 = \beta^2 \|g_n - g_m\|^2 + \|(\alpha I - T)(g_n - g_m)\|^2,$$

where $\mu = \alpha + i\beta$, α and β real. Since $\beta \neq 0$, we obtain the inequality $\epsilon > |\beta| \cdot \|g_n - g_m\|$, or $\|g_n - g_m\| < \epsilon/|\beta|$. Thus, the sequence of g's is Cauchy, and since H is complete, the g's converge to some vector, say h. We now observe that $Th - f = (Tg_n - f) + T(h - g_n) = (f_n - f) + T(h - g_n)$, and so $\|Th - f\| \leqslant \|f_n - f\| + \|T\| \cdot \|h - g_n\|$. Letting $n \to \infty$, we conclude that $Th = f$. Therefore, the mapping $\mu I - T$ is *onto*, and the proof is complete.

We now turn to the class of normal operators, which possess some of the important features of hermitian operators.

Definition 1: *An operator N is said to be normal if it commutes with its adjoint: $NN^* = N^*N$.*

A number of consequences of this definition are presented in Exercises 2 to 5.

Theorem 3: *The operator N is normal iff $\|Nf\| = \|N^*f\|$ for every vector f.*

PROOF: Suppose that N is normal. Then, for any vector f, $\|Nf\|^2 = (Nf, Nf) = (N^*Nf, f) = (NN^*f, f) = (N^*f, N^*f) = \|N^*f\|^2$, and so $\|Nf\| = \|N^*f\|$. Conversely, if $\|Nf\| = \|N^*f\|$ for every vector f, we obtain $((NN^* - N^*N)f, f) = (NN^*f, f) - (N^*Nf, f) = (N^*f, N^*f) - (Nf, Nf) = \|N^*f\|^2 - \|Nf\|^2 = 0$. By Theorem 5–1, $NN^* - N^*N = O$, or $NN^* = N^*N$, and so N is normal.

Theorem 4: *If g is an eigenvector of the normal operator N associated with the eigenvalue μ, then g is also an eigenvector of N^* associated with the eigenvalue $\bar{\mu}$.*

PROOF: According to the last part of Exercise 3, $N - \mu I$ is normal. Employing Theorem 3 we obtain $0 = \|(N - \mu I)g\| = \|(N - \mu I)^*g\| = \|(N^* - \bar{\mu}I)g\|$, and so $(N^* - \bar{\mu}I)g = o$, or $N^*g = \bar{\mu}g$.

Theorem 5: *If g_1 and g_2 are eigenvectors of the normal operator N, corresponding to distinct eigenvalues μ_1 and μ_2, respectively, then $(g_1, g_2) = 0$.*

PROOF: Employing the result of the preceding theorem, we obtain $\mu_1(g_1, g_2) = (\mu_1 g_1, g_2) = (Ng_1, g_2) = (g_1, N^*g_2) = (g_1, \bar{\mu}_2 g_2) = \mu_2(g_1, g_2)$. Hence, $(\mu_1 - \mu_2)(g_1, g_2) = 0$, and so $(g_1, g_2) = 0$.

Definition 2: *An operator T is said to be norm-preserving if $\|Tf\| = \|f\|$ for every vector f. (Note that this condition guarantees that T is one-to-one.)*

Theorem 6: *(a) T is norm-preserving iff $T^*T = I$.*

(b) If T is norm-preserving, it also preserves all inner products—i.e., $(Tf, Tg) = (f, g)$ for every pair of vectors f, g.

PROOF: (*a*) If $T^*T = I$, then, for every vector f, $\|f\|^2 = (f,f) = (T^*Tf, f) = (Tf, Tf) = \|Tf\|^2$, and so $\|Tf\| = \|f\|$. Conversely, if T is norm-preserving, then, for every vector f, $((I - T^*T)f, f) = (f,f) - (T^*Tf, f) = (f,f) - (Tf, Tf) = \|f\|^2 - \|Tf\|^2 = 0$, and so, by Theorem 5–1, $T^*T = I$.

(*b*) If T is norm-preserving, then, for every pair of vectors f, g we obtain $(Tf, Tg) = (T^*Tf, g) = (If, g) = (f, g)$.

Definition 3: *An operator U is said to be unitary if it is both norm-preserving and invertible.* (*Cf.* Exercise 6.)

Theorem 7: *U is unitary iff $UU^* = U^*U = I$. (Hence, a unitary operator is necessarily normal.)*

PROOF: The proof follows trivially from the preceding theorem and definition. Note that U is unitary iff U^* is unitary; also, U is unitary iff $U^* = U^{-1}$.

Theorem 8: *The product of any (finite) number of unitary operators is unitary.*

PROOF: Left as Exercise 7.

Theorem 9: *The spectrum of a unitary operator U lies entirely on the unit circle, $\{\lambda \mid |\lambda| = 1\}$.*

PROOF: (*a*) Since $\|U\| = 1$, it follows directly from Theorem 7–4 that the spectrum of U is confined to the closed unit disc, $\{\lambda \mid |\lambda| \leqslant 1\}$.

(*b*) Let $|\mu| < 1$. Then for any non-zero vector f we have $\|Uf\| = \|f\| > \|\mu f\|$, and so $(\mu I - U)f \neq o$. Thus, $\mu I - U$ is one-to-one.

(*c*) Assuming, as in (*b*), that $|\mu| < 1$, we shall show that the range of $(\mu I - U)$ is dense in H. (Note the resemblance to the proof of Theorem 2.) If this were not so, there would exist a non-zero vector h such that $((\mu I - U)f, h) = 0$ for every vector f. Choosing for f the vector U^*h, we would obtain $(\mu U^*h, h) = (UU^*h, h) = (h, h)$, and by the Schwarz inequality we would obtain $\|h\|^2 \leqslant |\mu| \cdot \|U^*\| \cdot \|h\|^2$. Dividing by $\|h\|^2$ and recalling that $\|U^*\| = 1$, we would obtain $|\mu| \geqslant 1$, contradicting our hypothesis.

(*d*) Continuing to assume that $|\mu| < 1$, we can now extend the result of (*c*) to show that the range of $\mu I - U$ consists of all of H; the proof is very similar to that of part (*c*) of Theorem 2, and we leave the details to the reader as Exercise 8. Hence, $\mu I - U$ is invertible, and so the spectrum is confined entirely to the circumference of the unit disc.

In Appendix H we provide a brief introduction to a particularly important unitary operator, the Fourier transform on $L^2(R)$.

Exercises

1. Carry out the calculation which is needed in part (*c*) of the proof of Theorem 2. (Of course, the fact that T is hermitian must be exploited.)

2. Prove that any scalar multiple of a hermitian operator is normal, and that any scalar multiple of a normal operator is also normal.

3. Prove that the product of two normal operators is normal iff each of them commutes with the adjoint of the other; in fact, if one of the commutation relations holds, then so does the other. In particular, if N is normal and α is any scalar, then $N - \alpha I$ is also normal.

4. Prove that every operator T can be expressed, in a unique manner, as $A + iB$, where A and B are hermitian, and that T is normal iff A and B commute. (For obvious reasons, A and B are termed the real part and imaginary part of T, respectively.)

5. Prove that if N is normal the equality $\|N^2\| = \|N\|^2$ holds.

6. Give an example of an operator which is norm-preserving but not unitary.

7. Prove Theorem 8.

8. Work out the details of part (*d*) of the proof of Theorem 9.

9. (*a*) Let T be any operator whose spectrum does not contain the number i. Prove that $(T + iI)(T - iI)^{-1} = (T - iI)^{-1}(T + iI)$, so that we may speak unambiguously of the operator $(T + iI)/(T - iI)$.

   (*b*) Suppose that T is hermitian, so that its spectrum certainly does not contain i. Prove that $(T + iI)/(T - iI)$ is unitary.

CHAPTER 7

OPERATORS ON FINITE-DIMENSIONAL SPACES

Before continuing with the development of the general theory of operators, we shall devote the present chapter to the study of some parts of the theory of operators on finite-dimensional spaces. The reader is surely aware that there exists a close connection, in finite-dimensional spaces, between linear transformations and matrices. In fact, this connection was for a long time the central idea of the whole theory of linear algebra, but in recent decades the concept of the transformation has come to play the central role, while the matrix now plays a secondary role, of significance primarily as a computational tool.

Throughout this chapter we are dealing with a *complex* linear space V of finite dimension n. In §1 and §2 the concepts of norm and inner product play no role.

We assume that the reader is familiar with the elementary manipulations of matrices and also that he is acquainted with the elements of the theory of determinants.

§1. MATRIX REPRESENTATION OF LINEAR TRANSFORMATIONS

Let T be a linear transformation on V and let the vectors $g_1, g_2, \ldots, g_n$ form a basis of V. Let Tg_i be represented, for each index i, in the form

$$Tg_i = \sum_{j=1}^{n} \alpha_{ji} g_j.$$

Then for any vector f we may write

$$f = \sum_{i=1}^{n} \gamma_i g_i$$

and

$$Tf = \sum_{i=1}^{n} \gamma_i T g_i = \sum_{i,j=1}^{n} \gamma_i \alpha_{ji} g_j = \sum_{j=1}^{n} \left\{ \sum_{i=1}^{n} \gamma_i \alpha_{ji} \right\} g_j.$$

Identifying f with the column vector

$$\begin{pmatrix} \gamma_1 \\ \gamma_2 \\ \cdot \\ \cdot \\ \cdot \\ \gamma_3 \end{pmatrix},$$

we see that the preceding equations can be rewritten in the form

$$Tf = \begin{pmatrix} \alpha_{11} & \alpha_{12} & \cdots & \alpha_{1n} \\ \alpha_{21} & \alpha_{22} & \cdots & \alpha_{2n} \\ \cdot & \cdot & & \cdot \\ \cdot & \cdot & & \cdot \\ \cdot & \cdot & & \cdot \\ \alpha_{n1} & \alpha_{n2} & \cdots & \alpha_{nn} \end{pmatrix} \begin{pmatrix} \gamma_1 \\ \gamma_2 \\ \cdot \\ \cdot \\ \cdot \\ \gamma_n \end{pmatrix}.$$

Thus the n-by-n matrix appearing in the preceding equation represents the transformation T. However, it is extremely important to realize that this matrix (which we shall denote by the symbol T_g) represents T with reference to a particular basis, namely, that formed by the vectors $g_1, g_2, \ldots, g_n$; if a different basis is selected, the representative matrix will usually be quite different from T_g. (Cf. Exercise 1.) If instead of the basis $\{g_1, g_2, \ldots, g_n\}$ we select the basis $\{h_1, h_2, \ldots, h_n\}$ how is the representative matrix altered? Clearly, each of the h's can be represented in the form

$$h_k = \sum_{l=1}^{n} \beta_{kl} g_l, \tag{1–1}$$

and similarly each of the g's can be expressed in the form

$$g_l = \sum_{m=1}^{n} \beta'_{lm} h_m.$$

Substituting from the latter equation into the former, we obtain

$$h_k = \sum_{l,m=1}^{n} \beta_{kl}\beta'_{lm}h_m,$$

and so, since the h's are linearly independent, we conclude that

$$\sum_{l=1}^{m} \beta_{kl}\beta'_{lm} = \delta_{km},$$

where δ_{km}, the *Kronecker delta*, equals 1 when $k = m$ and 0 when $k \neq m$.

In matrix notation, this last equation may be written $bb' = I_n$, where b and b' are the matrices containing the scalars β_{ij} and β'_{ij}, respectively, in the i-th row and j-th column, while I_n is the matrix consisting of ones down the main diagonal and zeros elsewhere. Thus, the matrices b and b' are inverses of each other, and therefore their determinants are reciprocals of each other, so that neither determinant vanishes. (Conversely, if the g's constitute a basis then the vectors $h_1, h_2, \ldots, h_n$ defined by equation (1–1) constitute a basis iff the determinant of the matrix b does not vanish.)

Referring back to the beginning of this section, we obtain

$$f = \sum_{i,m=1}^{n} \gamma_i\beta'_{im}h_m, \qquad Tf = \sum_{i,j,m=1}^{n} \gamma_i\alpha_{ji}\beta'_{jm}h_m.$$

If we now express f directly in terms of the h's:

$$f = \sum_{i=1}^{n} \gamma'_i h_i$$

and then identify f with the column-vector

$$\begin{pmatrix} \gamma'_1 \\ \gamma'_2 \\ \cdot \\ \cdot \\ \cdot \\ \gamma'_n \end{pmatrix}$$

we immediately obtain

$$\begin{pmatrix} \gamma'_1 \\ \gamma'_2 \\ \cdot \\ \cdot \\ \cdot \\ \gamma'_n \end{pmatrix} = \tilde{\beta}' \begin{pmatrix} \gamma_1 \\ \gamma_2 \\ \cdot \\ \cdot \\ \cdot \\ \gamma_n \end{pmatrix}, \qquad Tf = \tilde{\beta}' T_g \begin{pmatrix} \gamma_1 \\ \gamma_2 \\ \cdot \\ \cdot \\ \cdot \\ \gamma_n \end{pmatrix}, \tag{1–2}$$

where $\tilde{\beta}'$ is the transpose of the matrix β'. From the first equation of (1–2) we obtain

$$\begin{pmatrix} \gamma_1 \\ \gamma_2 \\ \cdot \\ \cdot \\ \cdot \\ \gamma_n \end{pmatrix} = (\tilde{\beta}')^{-1} \begin{pmatrix} \gamma_1' \\ \gamma_2' \\ \cdot \\ \cdot \\ \cdot \\ \gamma_n' \end{pmatrix}.$$

Now, by substituting into the second equation of (1–2), we obtain

$$Tf = \tilde{\beta}' T_g (\tilde{\beta}')^{-1} \begin{pmatrix} \gamma_1 \\ \gamma_2 \\ \cdot \\ \cdot \\ \cdot \\ \gamma_n \end{pmatrix}.$$

Thus, we have obtained the desired relation between the representative matrices T_g and T_h, namely:

$$T_h = \tilde{\beta}' T_g (\tilde{\beta}')^{-1}. \tag{1–3}$$

It might be well to emphasize that the matrices T_g and T_h represent the *same* operator, T, with respect to different bases, while the matrices β and β' relate the two different bases under consideration but have nothing to do with T.

From (1–3) we immediately obtain $\det T_h = (\det \tilde{\beta}')(\det T_g) \times (\det (\tilde{\beta}')^{-1}) = (\det T_g)(\det (\tilde{\beta}'\tilde{\beta}'^{-1})) = (\det T_g)(\det I_n) = \det T_g$. Thus, although the *matrices* T_g and T_h are in general quite different, their determinants are equal; that is to say, the value of the determinant represents an intrinsic property of the transformation T, so that we may speak unambiguously of the quantity $\det T$.

We leave to the reader, as Exercise 2, the proof of the facts that if T_1 and T_2 are operators and δ_1 and δ_2 are scalars, then $(\delta_1 T_1 + \delta_2 T_2)_g = \delta_1 (T_1)_g + \delta_2 (T_2)_g$ and $(T_1 T_2)_g = (T_1)_g (T_2)_g$; also, if T^{-1} exists, then $(T^{-1})_g = (T_g)^{-1}$, and T^{-1} exists iff $\det T \neq 0$. These results, simple though they are, provide the essence of the proof that (once a particular basis is selected) there exists an isomorphism between the algebra of linear transformations on V and the algebra of n-by-n matrices.

Exercises

1. Show that the zero operator and the identity operator are always represented by the same matrices respectively, regardless of the choice of basis.

2. Prove the statements made in the final paragraph.

3. Let V be the space of polynomials of degree $\leqslant 4$ and let Tf be the first derivative of f for every vector f in V. Set up the matrix representing T with respect to the basis $\{1, x, x^2, x^3, x^4\}$.

§2. EIGENVALUES AND EIGENVECTORS

The definitions of the terms eigenvalue, eigenvector, and spectrum do not require the concept of a norm. Hence, we may employ these concepts in our present setting. We shall see that, in the finite-dimensional case, the spectrum consists exclusively of eigenvalues, and the existence of at least one eigenvalue (and hence the non-emptiness of the spectrum) will follow from the fundamental theorem of algebra.

The existence of a non-zero vector f satisfying the relation $Tf = \mu f$ is known, from the elements of linear algebra, to be equivalent to the vanishing of det $(T - \mu I)$. Choosing a basis $\{g_1, g_2, \ldots, g_n\}$ and setting up the matrix T_g as in the preceding section, we immediately see that the eigenvalues of T are the roots of the so-called *characteristic equation* associated with T:

$$\begin{vmatrix} \alpha_{11} - \mu & \alpha_{12} & \cdots & \alpha_{1n} \\ \alpha_{21} & \alpha_{22} - \mu & \cdots & \alpha_{2n} \\ \cdot & \cdot & & \cdot \\ \cdot & \cdot & & \cdot \\ \cdot & \cdot & & \cdot \\ \alpha_{n1} & \alpha_{n2} & \cdots & \alpha_{nn} - \mu \end{vmatrix} = 0.$$

The left side of this equation is clearly a polynomial (in μ) of degree n, with leading term $(-1)^n\mu^n$. By the fundamental theorem of algebra, this polynomial, which is called the *characteristic polynomial* of the matrix T_g, possesses exactly n zeroes if multiplicity is counted; thus, there may be anywhere between one and n distinct eigenvalues. (Note that this argument fails completely in the real field.) The usual case is that the n zeros of the characteristic polynomial are all distinct. When this occurs one can give a very simple description of the linear transformation T under consideration; when multiple zeros occur the situation becomes more complicated, and we shall discuss it only briefly.

Theorem 1: *Let $\mu_1, \mu_2, \ldots, \mu_k$ be distinct eigenvalues of T (clearly, $k \leqslant n$) and let $g_1, g_2, \ldots, g_k$ be eigenvectors associated respectively with these eigenvalues. Then the g's are linearly independent.*

PROOF: Suppose that the g's were linearly dependent. Then we could select fewer than k of them (say $g_1, g_2, \ldots, g_j$) which are linearly independent and such that each of the remaining g's is expressible *uniquely* as a combination of the aforementioned g's. In particular, we could write

$$g_k = \alpha_1 g_1 + \alpha_2 g_2 + \cdots + \alpha_j g_j.$$

It would then follow that

$$Tg_k = \alpha_1 \mu_1 g_1 + \alpha_2 \mu_2 g_2 + \cdots + \alpha_j \mu_j g_j.$$

Since $Tg_k = \mu_k g_k$, the latter equation becomes

$$\mu_k g_k = \alpha_1 \mu_1 g_1 + \alpha_2 \mu_2 g_2 + \cdots + \alpha_j \mu_j g_j.$$

Multiplying the first equation by μ_k and subtracting from the last, we obtain

$$o = \alpha_1(\mu_1 - \mu_k)g_1 + \alpha_2(\mu_2 - \mu_k)g_2 + \cdots + \alpha_j(\mu_j - \mu_k)g_j.$$

Since the g's appearing in the last equation are linearly independent, all the coefficients must vanish, and since, by hypothesis, $\mu_i - \mu_k \neq 0$ for $1 \leqslant i \leqslant j$, we conclude that $\alpha_1 = \alpha_2 = \cdots = \alpha_j = 0$, and so $g_k = o$, contradicting the fact that g_k is an eigenvector.

Corollary: *If T possesses n distinct eigenvalues $\mu_1, \mu_2, \ldots, \mu_n$ and if $g_1, g_2, \ldots, g_n$ are eigenvectors associated respectively with these eigenvalues, they constitute a basis of the space V.*

PROOF: According to the theorem, the g's constitute a set of n linearly independent vectors in an n-dimensional space, hence they constitute a basis.

Corollary: *If T possesses n distinct eigenvalues, the manifold of solutions of each equation $Tf = \mu_k f$ is one-dimensional; i.e., it is impossible to find linearly independent vectors f_1, f_2 such that $Tf_1 = \mu_k f_1$ and $Tf_2 = \mu_k f_2$.*

PROOF: From the preceding corollary it is evident that the collection of vectors consisting of f_1, f_2, and all the g's except g_k would be linearly independent; this is impossible, since an n-dimensional space cannot contain more than n linearly independent vectors.

The latter corollary suggests the following definition.

Definition 1: *An eigenvalue μ is said to be simple, or non-degenerate, if the equation $Tf = \mu f$ does not possess two linearly independent solutions. If μ is not simple, it is said to be degenerate, and the degree of degeneracy* *is equal to the maximum number of linear independent eigenvectors associated with μ.*

Thus, the second corollary asserts that if all the eigenvalues are distinct they are simple. As indicated previously, the situation becomes more complicated when at least one of the eigenvalues is a multiple root of the characteristic polynomial. To begin with a trivial example, we see that the characteristic equation associated with the identity transformation I assumes the form $(1 - \mu)^n = 0$, and so the only eigenvalue is 1. Since every non-zero vector is obviously an eigenvector, the eigenvalue is degenerate of degree n, and it is possible, as in the case that n distinct eigenvalues exist, to select a basis consisting entirely of eigenvectors.

This result suggests very strongly that it is always possible to select a basis consisting entirely of eigenvectors, whether or not repetitions occur among the roots of the characteristic equation. However, this is not so, as shown by the following example, in which for definiteness we take $n = 4$. Let T be represented, with respect to some basis $\{g_1, g_2, g_3, g_4\}$, by the matrix

$$T_g = \begin{pmatrix} 1 & 1 & 0 & 0 \\ 0 & 1 & 1 & 0 \\ 0 & 0 & 1 & 1 \\ 0 & 0 & 0 & 1 \end{pmatrix}.$$

The characteristic equation is obviously $(1 - \mu)^4 = 0$ (even though T is not the identity). Thus, any eigenvector, when represented as a column-vector, must satisfy the equality

$$\begin{pmatrix} \alpha_1 + \alpha_2 \\ \alpha_2 + \alpha_3 \\ \alpha_3 + \alpha_4 \\ \alpha_4 \end{pmatrix} = \begin{pmatrix} \alpha_1 \\ \alpha_2 \\ \alpha_3 \\ \alpha_4 \end{pmatrix}$$

We immediately see that $\alpha_2 = \alpha_3 = \alpha_4 = 0$, and so every eigenvector must be a (non-zero) multiple of the vector corresponding to the column-vector

$$\begin{pmatrix} 1 \\ 0 \\ 0 \\ 0 \end{pmatrix};$$

i.e., g_1 and its multiples are the only eigenvectors, and so the eigenvectors do not span the space.

We refer the reader to texts on linear algebra for a detailed analysis of operators whose characteristic equation possesses repeated roots.

Returning to the case that the eigenvectors are known to span the space, let us choose a basis $\{g_1, g_2, \ldots, g_n\}$ of eigenvectors and let us write down in the same order their associated eigenvalues, $\{\mu_1, \mu_2, \ldots, \mu_n\}$. (Remember that repetitions may occur among the μ's.) Then it is quite obvious that the matrix T_g representing T with respect to this basis must assume the form

$$T_g = \begin{pmatrix} \mu_1 & & & & \mathbf{O} \\ & \mu_2 & & & \\ & & \cdot & & \\ & & & \cdot & \\ & & & & \cdot \\ \mathbf{O} & & & & \mu_n \end{pmatrix}.$$

Conversely, it is evident that if T can be represented, with respect to *some* basis, by a diagonal matrix, then the basis *must* consist of eigenvectors and the diagonal entries of the matrix must coincide with the eigenvalues, each eigenvalue appearing a number of times equal to its degree of degeneracy.

This result can be restated as follows, if we take account of the work in §1 dealing with the effect of a change of basis on the representative matrix: The eigenvectors of T span V iff the matrix T_f, for arbitrary choice of a basis $\{f_1, f_2, \ldots, f_n\}$, has the property that there exists an (invertible) matrix C such that $C^{-1}T_fC$ is a diagonal matrix; we say that T_f is *diagonable* and that C *diagonalizes* T_f.

We now return to the characteristic equation associated with T. When the left side of this equation is worked out, we obtain (cf. Exercise 1)

$$(-1)^n\{\mu^n - (\alpha_{11} + \alpha_{22} + \cdots + \alpha_{nn})\mu^{n-1} + \cdots + (-1)^n \det T_g\} = 0.$$

Since the roots of this equation are the eigenvalues of T and since the leading coefficient, namely $(-1)^n$, is independent of the choice of basis, it follows that *all* the coefficients must be independent of the choice of basis. In particular, the sum of the diagonal elements, $\alpha_{11} + \alpha_{22} + \cdots + \alpha_{nn}$, which is termed the *trace* of the matrix T_g, must be independent of the choice of basis, and by observing the constant term of the left side we find once again that the determinant of the representative matrix depends only on T, not on the choice of the basis. Furthermore, we easily see that the trace must equal the sum of the eigenvalues, while the quantity $\det T_g$ must equal the product of the eigenvalues.

As the last topics in this section we shall present the Hamilton-Cayley theorem and an application of this theorem to nilpotent operators.

Given any n-by-n matrix

$$A = \begin{pmatrix} \alpha_{11} & \alpha_{12} & \cdots & \alpha_{1n} \\ \alpha_{21} & \alpha_{22} & \cdots & \alpha_{2n} \\ \cdot & \cdot & & \cdot \\ \cdot & \cdot & & \cdot \\ \cdot & \cdot & & \cdot \\ \alpha_{n1} & \alpha_{n2} & \cdots & \alpha_{nn} \end{pmatrix},$$

we may obviously speak of the characteristic polynomial associated with A and of the eigenvalues of A, even though no linear space V and linear transformation on V are under consideration.

Theorem 2: (Hamilton-Cayley): *Let the variable μ in the characteristic polynomial of the matrix A be replaced by A; the resulting matrix reduces to the n-by-n zero matrix O_n. (This theorem is often stated as follows: Every [square] matrix satisfies its own characteristic equation.)*

PROOF: First suppose that the eigenvalues $\mu_1, \mu_2, \ldots, \mu_n$ of A are all distinct. Choose any n-dimensional space V and any basis

$$\{g_1, g_2, \ldots, g_n\}$$

of V, and let T be that operator on V whose representative matrix, with respect to the basis $\{g_1, g_2, \ldots, g_n\}$, is A. Choose eigenvectors $f_1, f_2, \ldots, f_n$ associated, respectively, with the eigenvalues $\mu_1, \mu_2, \ldots, \mu_n$. Since the characteristic polynomial, $c(\mu)$, can be written in factored form as follows:

$$c(\mu) = (\mu_1 - \mu)(\mu_2 - \mu) \cdots (\mu_n - \mu),$$

we readily obtain

$$c(A) = (\mu_1 I - A)(\mu_2 I - A) \cdots (\mu_n I - A).$$

Since $(\mu_n I - A)f_n = o$, we see that $c(A)f_n = o$. Since the order of the factors may be rearranged at pleasure, we obtain the more general result that $c(A)f_k = o$, $1 \leqslant k \leqslant n$. By linearity, $c(A)$ annihilates every linear combination of the vectors $f_1, f_2, \ldots, f_n$. Since the f's constitute a basis, $c(A)$ must annihilate every vector in V. Thus, $c(A)$ must represent the zero operator (with respect to the original basis $\{g_1, g_2, \ldots, g_n\}$), and so $c(A)$ must be the zero matrix, O_n.

If the characteristic equation of A possesses repeated roots, we employ a continuity argument. It is not difficult to show that, given $\epsilon > 0$, there exists a matrix $\tilde{A}$ whose eigenvalues are distinct and whose elements differ from the corresponding elements of A by less than ϵ in absolute value. Thus, $\tilde{c}(\tilde{A}) = O_n$, where $\tilde{c}(\mu)$ is the characteristic polynomial associated with $\tilde{A}$. Choosing a sequence of ϵ's converging to zero, we obtain the desired result, namely $c(A) = O_n$.

The Hamilton-Cayley theorem enables us to compute easily A^n once the matrices $A, A^2, \ldots, A^{n-1}$ are known. Then A^{n+1} is evidently expressible as a combination of $A, A^2, \ldots,$ and A^n; replacing A^n by the expression already obtained, we conclude that A^{n+1} (and all higher powers of A) can be expressed as linear combinations of the matrices $I_n, A, A^2, \ldots, A^{n-1}$.

Recalling the definition of a nilpotent operator given in Chapter 6, we define a *nilpotent matrix* as a (non-zero) matrix satisfying the equation $A^k = O_n$ for some positive integer k (where, of course, A is n-by-n). Let μ denote any eigenvalue of a nilpotent matrix A, and, thinking of A as representing an operator T on some space V, let f be an eigenvector of T corresponding to μ. Then $Tf = \mu f$, $T^2 f = \mu T f = \mu^2 f, \ldots, T^k f = \mu^k f$. Since T^k is represented by the matrix A^k, it follows that T^k must be the zero operator. Hence, $\mu^k f = o$, and since $f \neq o$, it follows that μ^k, and hence μ, must vanish.

Now, conversely, suppose that A has no eigenvalues except zero. Then the characteristic polynomial of A must be $(-1)^n \mu^n$, and the Hamilton-Cayley theorem now assures us that $A^n = O$. We have thus obtained the following interesting characterization of nilpotent matrices.

Theorem 3: *The non-zero n-by-n matrix A is nilpotent iff all its eigenvalues are zero. Furthermore, it is not necessary to go beyond A^n in the sequence $A^2, A^3, \ldots$ before reaching the matrix O_n. (Cf. Exercise* 3.)

Exercises

1. Work out explicit expressions for all the coefficients appearing in the characteristic equation.

2. By employing the formulas developed in §1, demonstrate the invariance of the trace under change of basis, without any consideration of eigenvalues.

3. Construct a nilpotent n-by-n matrix A such that $A^{n-1} \neq O_n$.

§3. FINITE-DIMENSIONAL INNER-PRODUCT SPACES

We now turn to the case that the space V is provided with an inner product. We shall see that a very complete and simple description of hermitian operators can be worked out and that a substantial portion of the theory can be carried over to normal operators. Furthermore, in the case of hermitian operators the existence of eigenvalues and the fact that the eigenvectors span V will be demonstrated without resorting to either the fundamental theorem of algebra or to the representation of operators by matrices. We shall frequently employ, without specific reference, results obtained in the previous chapter. It must be kept in mind that the arguments employed here are, by their nature, confined to finite-dimensional spaces.

Theorem 1: *If T is hermitian, then at least one of the numbers $\|T\|$, $-\|T\|$ is an eigenvalue of T.*

PROOF: Since $\|T\| = \sup_{\|f\|=1} |(Tf, f)|$, it follows that either $\|T\| = \sup_{\|f\|=1} (Tf, f)$ or $-\|T\| = \inf_{\|f\|=1} (Tf, f)$. It suffices to consider the former case, for otherwise we may simply replace T by $-T$. Introducing the symbol μ for $\|T\|$, we then have $\mu = \|T\| = \sup_{\|f\|=1} (Tf, f)$. Since V is locally compact, the unit vectors form a compact subset of V. Since (Tf, f) depends continuously on f, it follows that there exists a unit vector g such that $(Tg, g) = \mu$. We then obtain

$$\|Tg - \mu g\|^2 = (Tg - \mu g, Tg - \mu g) = (Tg, Tg) - 2\mu(Tg, g) + \mu^2(g, g)$$
$$= \|Tg\|^2 - 2\mu^2 + \mu^2 \leqslant \|T\|^2 - \mu^2 = 0.$$

Hence, $\|Tg - \mu g\| = 0$, and so $Tg = \mu g$.

Thus, not only have we shown that $\|T\|$ (or $-\|T\|$) is an eigenvalue of T; we have also shown how, in principle, the norm of T and an eigenvector associated with $\pm \|T\|$ can be found, namely by seeking a unit vector which maximizes $|(Tf, f)|$. Conversely, if $Tg = \pm\mu g$, then $(Tg, g) = \pm\mu$ (assuming $\|g\| = 1$). We shall later exhibit some specific computations relating to this remarkable characterization of the largest (in absolute value) eigenvalue.

Having found the eigenvalue λ_1 $(= \pm\mu)$ and a corresponding unit eigenvector, which we denote g_1, let us now consider the behavior of T when it is restricted to operating on vectors orthogonal to g_1. If $(g_1, h) = 0$, then $0 = \lambda_1(g_1, h) = (\lambda_1 g_1, h) = (Tg_1, h) = (g_1, Th)$. Thus, if we denote the subspace consisting of all vectors orthogonal to g_1 as S_1, we see that T may be considered as an operator mapping S_1 into S_1; clearly, the restriction of T to S_1 is still hermitian, and so we may repeat the preceding reasoning (if we exclude the trivial case that V is one-dimensional), obtaining a (unit) eigenvector g_2 associated with an eigenvalue λ_2, where $|\lambda_2| = \sup |(Tf, f)|$, the supremum being taken over all unit vectors orthogonal to g_1. By the very nature of the procedure employed, it is clear that $|\lambda_2| \leqslant |\lambda_1|$. If $n > 2$, we may repeat this procedure so long as there exists a non-zero vector orthogonal to each of the eigenvectors which have already been obtained. Thus the procedure which we have described automatically comes to a halt when a set of n eigenvectors $\{g_1, g_2, \ldots, g_n\}$ and a corresponding set of n eigenvalues $\{\lambda_1, \lambda_2, \ldots, \lambda_n\}$ have been obtained, the eigenvalues satisfying the chain of inequalities $|\lambda_1| \geqslant |\lambda_2| \geqslant \cdots \geqslant |\lambda_n|$. Furthermore, since the n eigenvectors are orthonormal, they are linearly independent and, hence, form an orthonormal basis of V.

It is not difficult to see that if at each stage we maximize (Tf, f) instead of $|(Tf, f)|$ we shall obtain the same set of eigenvalues, but in order of diminishing algebraic (rather than absolute) value; similarly, by minimizing (Tf, f) at each stage we would obtain the eigenvalues in ascending algebraic order.

Having obtained the eigenvalues $\lambda_1, \lambda_2, \ldots, \lambda_n$ and the eigenvectors $g_1, g_2, \ldots, g_n$, we see that the matrix T_g representing T with respect to the basis $\{g_1, g_2, \ldots, g_n\}$ is given by

$$T_g = \begin{pmatrix} \lambda_1 & & & & \mathbf{O} \\ & \lambda_2 & & & \\ & & \cdot & & \\ & & & \cdot & \\ & & & & \cdot \\ \mathbf{O} & & & & \lambda_n \end{pmatrix}. \tag{3–1}$$

In contrast to equation (2–1), the diagonal elements of (3–1) are certainly real, and it is not necessary to *assume* in the present case that the eigenvectors of T span the space V; the preceding argument shows that this condition is certainly satisfied.

Now suppose that T is hermitian and that we represent T with respect to an *arbitrary* orthonormal basis $\{h_1, h_2, \ldots, h_n\}$ by the matrix

$$T_h = \begin{pmatrix} \eta_{11} & \eta_{12} & \cdots & \eta_{1n} \\ \eta_{21} & \eta_{22} & \cdots & \eta_{2n} \\ \cdot & \cdot & & \cdot \\ \cdot & \cdot & & \cdot \\ \cdot & \cdot & & \cdot \\ \eta_{n1} & \eta_{n2} & \cdots & \eta_{nn} \end{pmatrix}, \tag{3–2}$$

where the elements of this matrix are determined by the equations (cf. §1)

$$Th_i = \sum_{j=1}^{n} \eta_{ji} h_j.$$

Taking account of the orthonormality of the h's, we obtain the chain of equalities $(Th_i, h_k) = (\sum_{j=1}^{n} \eta_{ji} h_j, h_k) = \sum_{j=1}^{n} \eta_{ji}(h_j, h_k) = \sum_{j=1}^{n} \eta_{ji}\delta_{jk} = \eta_{ki}$. Similarly, $\eta_{ik} = (Th_k, h_i) = (h_k, Th_i) = \overline{(Th_i, h_k)}$, and so $\eta_{ik} = \bar{\eta}_{ki}$. Thus, the matrix T_h must have the property that each entry is the conjugate of its image in the main diagonal. Conversely, if the vectors $\{h_1, h_2, \ldots, h_n\}$ form an orthonormal basis and if T_h possesses the preceding property, it is evident that T must be hermitian. For this reason, a matrix possessing the aforementioned property is termed *hermitian*, even if no inner-product space is under consideration. It should be emphasized that if the operator T is hermitian but the chosen basis $\{h_1, h_2, \ldots, h_n\}$ is not orthonormal, the matrix T_h will in general *not* be hermitian.

An obvious, but interesting and important, consequence of the foregoing analysis of hermitian operators is the following result.

Theorem 2: *A hermitian operator (on a finite-dimensional inner-product space) is positive iff all its eigenvalues are non-negative.*

This characterization of the eigenvalues and eigenvectors as the solutions to certain maximization problems possesses one unsatisfactory feature. We refer to the fact that the eigenvalues $\lambda_2, \lambda_3, \ldots$ and the corresponding eigenvectors are defined in terms of the preceding eigenvectors; for example, g_3 is determined as a unit vector which provides the solution to a certain maximization problem among all unit vectors orthogonal to g_1 and g_2. (Incidentally, these remarks apply whether the eigenvalues are indexed in order of descending algebraic value or of descending absolute value.) We now indicate briefly how the eigenvalues and eigenvectors can be characterized directly, without reference to eigenvectors previously obtained. For definiteness we suppose that the eigenvalues are to be indexed according to descending algebraic value; the modifications to be made when the eigenvalues are indexed otherwise (descending absolute value, ascending algebraic value, ascending absolute value) will be quite obvious.

We begin by repeating the characterizations of λ_1 and g_1: $\lambda_1 = \max_{\|f\|=1} (Tf, f)$ and g_1 is chosen as any unit vector such that $(Tg_1, g_1) = \lambda_1$. Now let h be any vector and let $\lambda(h)$ denote the value of max (Tf, f) when f is subjected to the condition $(f, h) = 0$ in addition to the usual condition $\|f\| = 1$. If $h = o$, then clearly $\lambda(h) = \lambda_1$ (since the orthogonality condition provides no restriction on f in this trivial case). Setting aside this trivial case, we may assume that h is a unit vector (for otherwise h may be replaced by $(1/\|h\|)h$) and we may then write $h = \alpha g_1 + \beta g$, where $|\alpha|^2 + |\beta|^2 = 1$ and g is some unit vector orthogonal to g_1. The condition $(f, h) = 0$ now assumes the form $\bar{\alpha}(f, g_1) + \bar{\beta}(f, g) = 0$. If we denote (f, g_1) and (f, g) by γ_1 and γ, respectively, the conditions $\|f\| = 1$ and $(f, h) = 0$ assume the forms

$$|\gamma_1|^2 + |\gamma|^2 = 1, \qquad \bar{\alpha}\gamma_1 + \bar{\beta}\gamma = 0, \tag{3–3}$$

respectively. We also note that $f = \gamma_1 g_1 + \tilde{\gamma}\tilde{g}$, where $|\tilde{\gamma}| = |\gamma|$ and $\tilde{g}$, like g, is a unit vector orthogonal to g_1. Hence,

$$(Tf, f) = |\gamma_1|^2 (Tg_1, g_1) + |\tilde{\gamma}|^2 (T\tilde{g}, \tilde{g}) + \gamma_1\bar{\tilde{\gamma}}(Tg_1, \tilde{g}) + \tilde{\gamma}\bar{\gamma}_1(T\tilde{g}, g_1). \tag{3–4}$$

The last two terms on the right side vanish, for $(Tg_1, \tilde{g}) = \lambda_1(g_1, \tilde{g}) = 0$ and $(T\tilde{g}, g_1) = (\tilde{g}, Tg_1) = \lambda_1(\tilde{g}, g_1) = 0$. Recalling that $(\tilde{g}, g_1) = 0$, we now see immediately that, for fixed γ_1 and $\tilde{\gamma}$, the maximum of (Tf, f) is equal to $\lambda_1 |\gamma_1|^2 + \lambda_2 |\tilde{\gamma}|^2$, or $\lambda_1 |\gamma_1|^2 + \lambda_2 |\gamma|^2$, and is achieved by choosing for $\tilde{g}$ the eigenvector g_2. Taking account of the first part of (3–3), we obtain $\lambda_1 |\gamma_1|^2 + \lambda_2 |\gamma|^2 = (\lambda_1 - \lambda_2) |\gamma_1|^2 + \lambda_2$; since $\lambda_1 - \lambda_2$ is non-negative, the right side of the last equality is *minimized* by choosing for γ_1 the value zero, and hence imposing on γ the condition $|\gamma| = 1$. The second part of (3–3) now imposes on β the value zero, and this in turn imposes on α the condition $|\alpha| = 1$.

Thus, we have shown the following: The quantity $\max_{\|f\|=1,(f,h)=0} (Tf, f)$ is minimized by choosing for h the eigenvector g_1 (or any non-zero multiple of g_1), and with this choice of h the aforementioned maximum is λ_2. In other words, λ_2 is the smallest possible value of $\max_{\|f\|=1} (Tf, f)$ which can be obtained by imposing a restriction of the form $(f, h) = 0$. By a rather straightforward extension of the preceding argument we can show that λ_3 is the smallest possible value of $\max_{\|f\|=1} (Tf, f)$ which can be obtained by imposing on f *two* restrictions of the form $(f, h_1) = (f, h_2) = 0$. By repeating this argument, we obtain the following direct characterization of each of the eigenvalues $\lambda_1, \lambda_2, \ldots, \lambda_n$.

Theorem 3 (Minimax Theorem): *The eigenvalue λ_k $(1 \leq k \leq n)$ of the hermitian operator T is equal to the minimum value of the quantity $\max_{\|f\|=1} (Tf, f)$ which can be achieved by imposing on the vector f restrictions of the form $(f, h_1) = (f, h_2) = \cdots = (f, h_{k-1}) = 0$.*

Although the determination of eigenvalues (of a linear transformation on a finite-dimensional space) is in principle a problem of elementary algebra (the solution of a polynomial equation), a huge amount of literature exists on the problem of obtaining accurate approximations for the eigenvalues (and eigenvectors). In the remainder of this section and in the following section we shall discuss briefly two methods, both of which are restricted to hermitian operators.

Let the hermitian operator T be represented, as before, with respect to some orthonormal basis $\{h_1, h_2, \ldots, h_n\}$ by the matrix (3–2). As we have seen, this matrix must be hermitian. It is quite clear that we may consider V, the space on which T operates, to be l_n^2 (with the usual definition of inner product). If we write the typical vector g of V in the form

$$g = \begin{pmatrix} \gamma_1 \\ \gamma_2 \\ \cdot \\ \cdot \\ \cdot \\ \gamma_n \end{pmatrix},$$

it is readily seen that

$$(Tg, g) = \sum_{i,j=1}^{n} \eta_{ij}\gamma_i\bar{\gamma}_j.$$

Thus, if we enumerate the eigenvalues in descending algebraic order, we obtain

$$\lambda_1 = \max \sum_{i,j=1}^{n} \eta_{ij}\gamma_i\bar{\gamma}_j, \qquad \lambda_n = \min \sum_{i,j=1}^{n} \eta_{ij}\gamma_i\bar{\gamma}_j,$$

the max and min being taken subject to the constraint

$$\sum_{i=1}^{n} |\gamma_i|^2 = 1.$$

In particular, if we choose $\gamma_k = \delta_{km}$ for some index m, we obtain $\lambda_n \leqslant \eta_{mm} \leqslant \lambda_1$. Since these inequalities hold for every choice of m, we obtain

$$\lambda_n \leqslant \min_{1 \leqslant m \leqslant n} \eta_{mm} \leqslant \max_{1 \leqslant m \leqslant n} \eta_{mm} \leqslant \lambda_1.$$

A more ambitious procedure consists in choosing a large number of unit vectors, spread as nearly densely as possible (this phrase, of course, is very imprecise) over the set $\{g \mid \|g\| = 1\}$ and evaluating (Tg, g) for each of these vectors. We illustrate this idea in the case of an operator T having, with respect to some orthonormal basis, the representative matrix

$$\begin{pmatrix} 1 & -3 \\ -3 & 2 \end{pmatrix}.$$

Let us choose the following unit vectors:

$$g_1 = \begin{pmatrix} 1 \\ 0 \end{pmatrix}, \quad g_2 = \begin{pmatrix} \frac{1}{2}\sqrt{3} \\ \frac{1}{2} \end{pmatrix}, \quad g_3 = \begin{pmatrix} 0.8 \\ 0.6 \end{pmatrix}, \quad g_4 = \begin{pmatrix} \frac{1}{2}\sqrt{2} \\ \frac{1}{2}\sqrt{2} \end{pmatrix},$$

$$g_5 = \begin{pmatrix} 0.6 \\ 0.8 \end{pmatrix}, \quad g_6 = \begin{pmatrix} \frac{1}{2} \\ \frac{1}{2}\sqrt{3} \end{pmatrix}, \quad g_7 = \begin{pmatrix} 0 \\ 1 \end{pmatrix}, \quad g_8 = \begin{pmatrix} -\frac{1}{2} \\ \frac{1}{2}\sqrt{3} \end{pmatrix},$$

$$g_9 = \begin{pmatrix} -0.6 \\ 0.8 \end{pmatrix}, \quad g_{10} = \begin{pmatrix} -\frac{1}{2}\sqrt{2} \\ \frac{1}{2}\sqrt{2} \end{pmatrix}, \quad g_{11} = \begin{pmatrix} -0.8 \\ 0.6 \end{pmatrix}, \quad g_{12} = \begin{pmatrix} -\frac{1}{2}\sqrt{3} \\ \frac{1}{2} \end{pmatrix}.$$

(Note that we are confining ourselves to trial vectors with *real* components; the justification of this step is left as Exercise 1. Also, note that if g_k is any one of the trial vectors, the vector $-g_k$ is not employed, for $(Tg_k, g_k) = (T(-g_k), -g_k)$, and so no additional information would be obtained.) The numbers (Tg_k, g_k) vary between -1.52 and 4.52. Thus, $\lambda_2 \leqslant -1.52$ and $\lambda_1 \geqslant 4.52$.

Of course, in this particular case the explicit determination of the eigenvalues is completely trivial. We easily obtain

$$\det(T - \lambda I) = \lambda^2 - 3\lambda - 7,$$

and so

$$\lambda_1 = \frac{3 + \sqrt{37}}{2} = 4.541 \cdots, \qquad \lambda_2 = \frac{3 - \sqrt{37}}{2} = -1.541 \cdots.$$

This computation provides a rather crude illustration of the Rayleigh-Ritz procedure, which will be considered in a little more detail when we deal in the next chapter with infinite-dimensional spaces, particularly in connection with the estimation of eigenvalues of symmetric kernels.

Exercises

1. Let T be a hermitian operator on a complex n-dimensional vector space, and suppose that, with respect to some orthonormal basis $\{h_1, h_2, h_3, \ldots, h_n\}$, the elements of the representative matrix T_h are all real. Show that it is possible to find a basis of eigenvectors whose components, with respect to the basis $\{h_1, h_2, \ldots, h_n\}$, are all real.

2. Suppose that, in addition to the assumptions made in Exercise 1, the elements of T_h are all strictly positive. Prove that $\|T\|$ is an eigenvalue, that it is non-degenerate, that associated with this eigenvalue there exists an eigenvector whose components with respect to the basis $\{h_1, h_2, \ldots, h_n\}$ are all positive, and that $-\|T\|$ is *not* an eigenvalue of T.

3. Let A be any hermitian matrix, with eigenvalues $\lambda_1, \lambda_2, \ldots, \lambda_n$ (in descending algebraic order). Prove that if any one of the diagonal elements (which are, of course, all real) is increased while all other elements are left fixed, then the new set of eigenvalues $\tilde{\lambda}_1, \tilde{\lambda}_2, \ldots, \tilde{\lambda}_n$ (also in descending algebraic order) are such that each of the n inequalities $\tilde{\lambda}_k \geqslant \lambda_k$ must hold, with strict inequality for at least one value of k.

§4. KELLOGG'S METHOD OF ESTIMATING THE LARGEST EIGENVALUE

For ease in exposition we first suppose that the hermitian operator T is known to be positive and that the largest eigenvalue is known to be non-degenerate. However, it will then be shown that with minor modifications the two restrictions just imposed may be dropped. However, the condition that T should be hermitian cannot be omitted.

From the work of the preceding section we know that the eigenvalues of T are all non-negative and that, if they are denoted in descending algebraic order (which, under our hypothesis that T is positive, also implies descending absolute value) by $\lambda_1, \lambda_2, \ldots, \lambda_n$, with corresponding unit eigenvectors $g_1, g_2, \ldots, g_n$, then for any vector f the equalities

$$f = \sum_{k=1}^{n} (f, g_k) g_k, \qquad Tf = \sum_{k=1}^{n} \lambda_k (f, g_k) g_k$$

hold. In fact, more generally, for any positive integer m, we obtain the equality

$$T^m f = \sum_{k=1}^{n} \lambda_k^m (f, g_k) g_k.$$

Assuming that $(f, g_1) \neq 0$, we obtain

$$(T^m f, f) = \sum_{k=1}^{n} \lambda_k^m |(f, g_k)|^2 = \lambda_1^m |(f, g_1)|^2 \left\{ 1 + \sum_{k=2}^{n} \left(\frac{\lambda_k}{\lambda_1} \right)^m \frac{|(f, g_k)|^2}{|(f, g_1)|^2} \right\}.$$

Replacing m by $m + 1$ and dividing, we obtain

$$\frac{(T^{m+1} f, f)}{(T^m f, f)} = \lambda_1 \cdot \frac{1 + \cdots}{1 + \cdots},$$

where the quantities denoted by $\cdots$ approach zero with increasing m (since the fractions λ_k/λ_1 are each strictly less than unity).

Hence, we have shown that λ_1, the largest eigenvalue, is given by the limiting relation

$$\lambda_1 = \lim_{m \to \infty} \frac{(T^{m+1} f, f)}{(T^m f, f)},$$

where f is any vector which is not orthogonal to g_1. (Of course, since g_1 is itself unknown, there is a possibility of making an unfortunate choice for f, but in some cases [cf. Exercise 1] it is possible to guard against this occurrence.) Obviously, the more closely the direction of f agrees with that of g_1, the faster is the convergence of the sequence of fractions $(T^{m+1}f, f)/(T^m f, f)$ to λ_1.

Now we shall show that the procedure just described furnishes the eigenvector g_1 as well as the eigenvalue λ_1. Clearly,

$$\|T^m f\| = \left\{ \sum_{k=1}^{n} \lambda_k^{2m} |(f, g_k)|^2 \right\}^{1/2} = \lambda_1^m |(f, g_1)| \{1 + \cdots\},$$

where, as before, $\cdots$ denotes an expression which approaches zero with increasing m. Hence,

$$\frac{T^m f}{\|T^m f\|} = \frac{(f, g_1)}{|(f, g_1)|} g_1 + \cdots.$$

Since $(f, g_1)/|(f, g_1)|$ is a constant of unit modulus, we may assume that it is equal to unity, for this merely amounts to replacing g_1 by another unit eigenvector associated with λ_1. Thus, we obtain the following remarkable formula for the unit eigenvector (or, more precisely, *a* unit eigenvector) associated with λ_1:

$$g_1 = \lim_{m\to\infty} \frac{T^m f}{\|T^m f\|}\,.$$

We now comment briefly on the possibility of extending the scope of the Kellogg procedure. First, it is not difficult to see, by a trivial change in the reasoning, that the method is effective even if the largest eigenvalue λ_1 is degenerate; all that is necessary is that the trial vector f should not be orthogonal to the manifold spanned by the eigenvectors associated with λ_1.

Secondly, even if T is not positive, T^2 *is* positive, with the same eigenvectors as T and with eigenvalues which are the squares of those of T. If T^2 has a non-degenerate largest eigenvalue, then it is evident that the *largest* or *smallest* eigenvalue of T—the largest in absolute value—is also non-degenerate, and the procedure described previously leads to this eigenvalue and to the corresponding eigenvector. However, complications may arise if the largest eigenvalue of T^2 is degenerate, in particular if the largest and smallest eigenvalues of T are the negatives of each other. However, in Exercise 2 we indicate how this complication may be avoided.

Returning to the case that T is known to be positive, let us suppose that f is any vector such that $Tf \neq o$. Then, by the generalized Schwarz inequality, which, we recall, asserts that $|(Tf, g)|^2 \leqslant (Tf, f)(Tg, g)$ for any positive operator T and any vectors f and g, it follows that $(Tf, f) > 0$. (Cf. Exercise 5–6 of Chapter 6.) Since $(T^2 f, f) = (Tf, Tf) > 0$, it follows that $T^2 f = T(Tf) \neq o$, and so, by repeating the preceding reasoning, we conclude that $(T(Tf), Tf) = (T^2 f, Tf) = (T^3 f, f) > 0$. Repeating this argument indefinitely, we conclude that $(T^k f, f) > 0$ for $k = 0, 1, 2, \ldots$. Now, in the generalized Schwarz inequality replace f by $T^{(m-2)/2} f$ and g by $T^{(m/2)} f$, for $m \geqslant 2$. (Of course, when m is odd, $T^{(m-2)/2}$ and $T^{m/2}$ are understood to signify $T^{(m-3)/2} T^{1/2}$ and $T^{(m-1)/2} T^{1/2}$, respectively, where $T^{1/2}$ is the positive square root of T, whose existence and uniqueness have been demonstrated in the previous chapter.) We then obtain $|(TT^{(m-2)/2} f,\ T^{m/2} f)|^2 \leqslant (TT^{(m-2)/2} f,\ T^{(m-2)/2} f)(TT^{m/2} f,\ T^{m/2} f)$. Taking account of the hermitian character of all the operators involved, we obtain

$$(T^m f, f)^2 \leqslant (T^{m-1} f, f)(T^{m+1} f, f),$$

or

$$\frac{(T^m f, f)}{(T^{m-1} f, f)} \leqslant \frac{(T^{m+1} f, f)}{(T^m f, f)}$$

and so the ratios $(T^{m+1} f, f)/(T^m f, f)$ which appear in the Kellogg procedure must *increase* with m to the largest eigenvalue. Of course, this result does not hold in general if T is not positive.

Exercises

1. Let T be represented, with respect to a certain orthonormal basis $\{h_1, h_2, h_3\}$, by the matrix

$$T_h = \begin{pmatrix} 1 & 2 & 3 \\ 2 & 2 & 4 \\ 3 & 4 & 5 \end{pmatrix}.$$

   Show that the vector $f = h_1 + h_2 + h_3$ may be safely used as a trial vector in the Kellogg procedure. With this choice of f, carry out several stages of the Kellogg procedure for determining both λ_1 and g_1. (Hint: Refer to Exercise 3–2.)

2. Let T be hermitian with eigenvalues $\lambda_1, \lambda_2, \ldots, \lambda_n$. Show that if c is any real number, the operator $T + cI$ is also hermitian, with the same eigenvectors as T and with eigenvalues $\lambda_1 + c$, $\lambda_2 + c, \ldots, \lambda_n + c$. Show how this fact may be used to obtain a monotone sequence of approximations for λ_1 (and also for λ_n).

§5. SPECTRAL REPRESENTATION OF HERMITIAN OPERATORS

We shall present in this section the so-called spectral theorem for a hermitian operator. It is, in fact, perfectly trivial (in the finite-dimensional case, to which we are restricting attention in the present chapter), but it turns out to have a natural generalization which constitutes the major theorem concerning hermitian operators on infinite-dimensional Hilbert spaces. We shall allude briefly to this generalization in the following chapter.

For convenience we now reverse our previous indexing of the eigenvalues of the hermitian operator T; we now enumerate them so that $\lambda_1 \leqslant \lambda_2 \leqslant \cdots \leqslant \lambda_n$. As before, we let the unit vector g_k denote an eigenvector associated with λ_k. (Of course, if any eigenvalue is degenerate the collection of eigenvectors associated with this eigenvalue is understood to be orthonormal. The reader may find it helpful on first reading to confine attention to the case that the eigenvalues are all simple and then observe the minor changes in reasoning needed to cover the general case.) For each real number λ, let each of the eigenvalues $\{\lambda_1, \lambda_2, \ldots, \lambda_k\}$ be $\leqslant\lambda$ while the remaining eigenvalues exceed λ. (Of course, $k = k(\lambda)$; if $\lambda < \lambda_1$, the set $\{\lambda_1, \lambda_2, \ldots, \lambda_k\}$ is empty, while if $\lambda \geqslant \lambda_n$ the equality $k = n$ holds.) Let S_λ denote the subspace spanned by the vectors $g_1, g_2, \ldots, g_k$ (so that $S_\lambda \subseteq S_\mu$ if $\lambda \leqslant \mu$), and let P_λ denote the projection operator with range S_λ.

We make the following observations about the family $\{P_\lambda\}$: (*a*) $P_\lambda = O$ if $\lambda < \lambda_1$; (*b*) $P_\lambda = I$ if $\lambda \geqslant \lambda_n$; (*c*) $P_\lambda P_\mu = P_\mu P_\lambda = P_{\min\{\lambda,\mu\}}$; (*d*) if $\lambda \leqslant \mu$, $P_\mu - P_\lambda$ is a projection; (*e*) if λ is *not* an eigenvalue of T, then P_μ remains constant when μ varies over a sufficiently small interval containing λ; and (*f*) if λ *is* an eigenvalue of T, then, for all sufficiently small *positive* ϵ, $P_{\lambda+\epsilon} = P_\lambda$, but $P_{\lambda-\epsilon} \neq P_\lambda$; in fact, $P_{\lambda+\epsilon} - P_{\lambda-\epsilon}$ is the projection upon the subspace spanned by the g's associated with the eigenvalue λ.

Roughly speaking, as λ traverses R from $-\infty$ to $+\infty$ the projection P_λ grows from O to I, the growth occurring at the eigenvalues of T.

The family $\{P_\lambda\}$ (i.e., the function sending λ into P_λ) is called the *resolution of the identity* associated with T.

Examples

(*a*) If $T = O$, $P_\lambda = O$ for $\lambda < 0$, $P_\lambda = I$ for $\lambda \geqslant 0$.

(*b*) If $T = I$, $P_\lambda = O$ for $\lambda < 1$, $P_\lambda = I$ for $\lambda \geqslant 1$.

(*c*) If T is a projection other than O and I, then $P_\lambda = O$ for $\lambda < 0$, $P_\lambda = I - T$ for $0 \leqslant \lambda < 1$, $P_\lambda = I$ for $\lambda \geqslant 1$.

Now we show how the operator T can be expressed, or synthesized, in terms of the corresponding resolution of the identity. For simplicity of exposition, let us assume that the eigenvalues are all non-degenerate, so that $\lambda_1 < \lambda_2 < \cdots < \lambda_n$. Since the eigenvectors $g_1, g_2, \ldots, g_n$ are orthonormal and span the space, we know that for every vector f the following equalities must hold:

$$f = \sum_{k=1}^{n} (f, g_k) g_k, \tag{5–1}$$

$$Tf = \sum_{k=1}^{n} (f, g_k) T g_k = \sum_{k=1}^{n} \lambda_k (f, g_k) g_k. \tag{5–2}$$

Now, let us choose numbers $\mu_0, \mu_1, \ldots, \mu_n$ such that $\mu_0 < \lambda_1 < \mu_1 < \lambda_2 < \cdots < \mu_{n-1} < \lambda_n < \mu_n$. Then clearly, for $1 \leqslant k \leqslant n$, $P_{\mu_k} - P_{\mu_{k-1}}$ is the projection on the one-dimensional manifold spanned by g_k, and $(P_{\mu_k} - P_{\mu_{k-1}})f = (f, g_k)g_k$.

Hence, (5–2) may be rewritten as follows:

$$Tf = \sum_{k=1}^{n} \lambda_k (P_{\mu_k} - P_{\mu_{k-1}}) f. \tag{5–3}$$

Thus, dropping f from both sides, we obtain the following remarkable representation of T:

$$T = \sum_{k=1}^{n} \lambda_k (P_{\mu_k} - P_{\mu_{k-1}}). \tag{5–4}$$

It is helpful to replace $P_{\mu_k} - P_{\mu_{k-1}}$ by the symbol $\delta P(\lambda_k)$, meaning the *growth* of P_λ as λ crosses the number λ_k (from left to right). Then we obtain

$$T = \sum_{k=1}^{n} \lambda_k \, \delta P(\lambda_k). \tag{5–5}$$

This is known as the *spectral representation* of *T;* we have proven that every *hermitian* operator on a *finite-dimensional* Hilbert space possesses a spectral representation, involving the eigenvalues of the operator and the resolution of the identity associated with the operator. This result is known as the *spectral theorem* (for finite-dimensional spaces). (The reader should have no difficulty in seeing that (5–5) remains correct even if T possesses one or more degenerate eigenvalues.)

The right side of (5–5) will (hopefully) suggest to the reader that an integration process of some sort is lurking in the background. If the reader has studied the (Riemann-)Stieltjes integral he should have no difficulty in seeing that the equality

$$(Tf, g) = \sum_{k=1}^{n} \lambda_k (\delta P(\lambda_k) f, g), \tag{5–6}$$

which follows immediately from (5–5), can also be written in the form

$$(Tf, g) = \int_{-\infty}^{\infty} \lambda \, d(P_\lambda f, g). \tag{5–7}$$

(For the benefit of the reader who has not yet encountered the Stieltjes integral, we provide a very brief account in Appendix *C*.) Hence, dropping the inner-product symbol in (5–7) and the vectors f and g, we are led to rewrite (5–5) in the form

$$T = \int_{-\infty}^{\infty} \lambda \, dP_\lambda. \tag{5–8}$$

Exercises

1. Let the operator T possess, with respect to a certain orthonormal basis $\{h_1, h_2, h_3\}$, the representative matrix

$$T_h = \begin{pmatrix} 1 & 1 & 1 \\ 1 & 1 & 1 \\ 1 & 1 & 1 \end{pmatrix}.$$

   Work out the resolution of the identity associated with T.

2. Let T be any hermitian operator and let p denote any polynomial in one variable (complex coefficients permitted). Show that, in analogy with (5–7), the equality

$$(p(T)f, g) = \int_{-\infty}^{\infty} p(\lambda)\, d(P_\lambda f, g)$$

   holds, and that $p(T)$ is hermitian iff $p(\lambda)$ is real when λ is an eigenvalue of T. (In particular, if p possesses real coefficients, this condition is satisfied.)

3. As in Exercise 2, let T be hermitian, and let p and q be two polynomials. Show that $p(T) = q(T)$ iff $p(\lambda) = q(\lambda)$ whenever λ is an eigenvalue of T.

§6. SPECTRAL REPRESENTATION OF NORMAL OPERATORS

In the previous chapter we have seen that normal operators share with hermitian operators the property that eigenvectors associated with distinct eigenvalues are orthogonal. Thus, if the normal operator N defined on a finite-dimensional space has no degenerate eigenvalues, it is easily seen that the eigenvectors form an orthogonal basis, and it is not difficult to prove that even when degeneracy occurs it is still possible to choose an orthonormal basis consisting entirely of eigenvectors. (Cf. Exercise 1.)

Recalling that every normal operator N can be expressed in the form $A + iB$, where A and B are hermitian operators which commute with each other, let the eigenvectors $g_1, g_2, \ldots, g_n$ of N form an orthonormal basis, with corresponding eigenvalues $\lambda_1, \lambda_2, \ldots, \lambda_n$. Since the λ_k's are, in general, non-real, we write

$$\lambda_k = \alpha_k + i\beta_k, \qquad \alpha_k \text{ and } \beta_k \text{ real.}$$

Then we know that $N^*g_k = \bar{\lambda}_k g_k$, and it is then readily seen that g_k is an eigenvector of both A and B:

$$Ag_k = \alpha_k g_k, \qquad Bg_k = \beta_k g_k.$$

Thus, taking account of the results of the previous section we see that we may associate with A and B, respectively, resolutions of the identity P_λ and Q_λ:

$$A = \int_{-\infty}^{\infty} \lambda \, dP_\lambda, \qquad B = \int_{-\infty}^{\infty} \lambda \, dQ_\lambda.$$

It should be quite obvious, from the fact that P_λ and Q_λ are projections on subspaces spanned by none, some, or all of the vectors $g_1, g_2, \ldots, g_n$, that the equality

$$(P_{\mu_1} - P_{\mu_2})(Q_{\nu_1} - Q_{\nu_2}) = (Q_{\nu_1} - Q_{\nu_2})(P_{\mu_1} - P_{\mu_2})$$

must hold whenever $\mu_1 \geqslant \mu_2$ and $\nu_1 \geqslant \nu_2$, and so the operator defined by each side of the preceding equation is itself a projection. For any pair of vectors f, g it is now easy, although perhaps a little tedious, to justify the following analogues of (5–7) and (5–8):

$$(Nf, g) = \int_{-\infty}^{\infty}\int_{-\infty}^{\infty} (\mu + i\nu) \, d(P_\mu f, g) \, d(Q_\nu f, g),$$

$$N = \int_{-\infty}^{\infty}\int_{-\infty}^{\infty} (\mu + i\nu) \, dP_\mu \, dQ_\nu.$$

It is customary to set $\mu + i\nu$ equal to the complex variable z and to define the projection E_z as the product $P_\mu Q_\nu$ $(= Q_\nu P_\mu)$. The preceding formulas are then rewritten as follows:

$$(Nf, g) = \iint_{\substack{\text{complex} \\ \text{plane}}} z \, d(E_z f, g),$$

$$N = \iint_{\substack{\text{complex} \\ \text{plane}}} z \, dE_z.$$

Of course, E_z is termed the resolution of the identity associated with N.

Exercises

1. Show that an operator N is normal iff it can be represented, with respect to some orthonormal basis, by a diagonal matrix.

2. Show that an operator U is unitary iff it can be represented, with respect to some orthonormal basis, by a diagonal matrix whose diagonal entries are all of absolute value one.

3. Prove that every normal operator N can be expressed in a unique manner as the product of a positive operator and a unitary operator and that the latter two operators commute.

4. (*a*) Let T be hermitian, with resolution of the identity P_λ, and let M be any subset of the real number system R. Give a reasonable interpretation of $\int_M \lambda \, dP_\lambda$.
   (*b*) Similarly, let N be normal, with resolution of the identity E_z, and let M be any subset of the complex number system $\mathfrak{C}$. Give a reasonable interpretation of $\iint_M z \, dE_z$.

5. Let U be a unitary operator. Define, on the interval $[0, 2\pi]$, an increasing family of projections $\{P_\theta\}$ (i.e., $P_{\theta_1} \leqslant P_{\theta_2}$ if $\theta_1 \leqslant \theta_2$) such that $P_0 = O$, $P_{2\pi} = I$, and, with a reasonable interpretation of the integral,

$$U = \int_0^{2\pi} e^{i\theta} \, dP_\theta.$$

CHAPTER 8

ELEMENTS OF SPECTRAL THEORY IN INFINITE-DIMENSIONAL HILBERT SPACES

The generalization to infinite-dimensional spaces of the theory presented in the previous chapter is in a very incomplete state of development. In this chapter we shall give a brief presentation of some of the most important ideas that have emerged during the development of the theory until the present. Most of our exposition will refer to Hilbert spaces, for which the present state of the theory, incomplete though it may be, is much more advanced than for Banach spaces.

§1. COMPLETELY CONTINUOUS OPERATORS

The concept of complete continuity plays an important role in the theory of operators and in the application of the theory to many problems of hard (as distinguished from soft, or abstract) analysis.

Definition 1: *An operator T on a normed linear space is said to be "completely continuous" (the terms "compact" and "totally bounded" are also used) if for every bounded sequence of vectors $f_1, f_2, f_3, \ldots$ the corresponding sequence $Tf_1, Tf_2, Tf_3, \ldots$ contains a subsequence which is convergent.*

Examples

(*a*) Every operator on a finite-dimensional normed linear space is completely continuous. (While this fact should be immediately obvious, it is also guaranteed by the corollary to Theorem 3.)

(*b*) The zero operator on any normed linear space is obviously completely continuous.

(*c*) If S is a subspace of an infinite-dimensional Hilbert space, then the projection on S, P_S, is completely continuous iff S is finite-dimensional. (Cf. Exercise 1.) In particular, it follows that the identity operator on an infinite-dimensional Hilbert space is not completely continuous. (Cf. Exercise 2.)

(*d*) An especially important example of a completely continuous operator is furnished by an integral transform. Consider the (real or complex) Banach space $C([a, b])$, where a and b are finite. (Of course, it is understood that the maximum norm is employed.) Let K (for *kernel*) be a continuous scalar-valued function defined on the square $[a, b] \times [a, b]$, and for each member f of $C([a, b])$ let $Tf = g$, where

$$g(x) = \int_a^b K(x, y) f(y)\, dy.$$

Obviously $Tf \in C([a, b])$; in fact, this would still be true under the weaker assumption that $f \in L^1([a, b])$. Furthermore, T is evidently linear, and the trivial inequality

$$|Tf(x)| \leqslant \{\max |K(x, y)|\} \cdot (b - a) \cdot \|f\|$$

shows that T is bounded, with norm not exceeding $\{\max |K(x, y)|\} \cdot (b - a)$. Thus, T is an operator.

Next, choose any two values of x in $[a, b]$, say x_1 and x_2. Then

$$\begin{aligned} |Tf(x_2) - Tf(x_1)| &\leqslant \int_a^b |K(x_2, y) - K(x_1, y)| \cdot |f(y)|\, dy \\ &\leqslant (b - a) \cdot \|f\| \cdot \max_{a \leqslant y \leqslant b} |K(x_2, y) - K(x_1, y)|. \end{aligned}$$

Since K is continuous on a compact subset of the plane, we can, given any $\delta > 0$, find a positive number η (depending only on δ and on the given kernel K) such that $|K(x_2, y) - K(x_1, y)| < \delta$ whenever $|x_2 - x_1| < \eta$. Now let positive numbers ϵ and A be given, let f be subjected to the restriction $\|f\| \leqslant A$, let $\delta = \epsilon/((b - a)A)$, and then choose η corresponding to this choice of δ. Then, whenever $|x_2 - x_1| < \eta$, the inequality $|Tf(x_2) - Tf(x_1)| < (b - a)A\delta = \epsilon$ must hold. Hence, the family of all vectors $\{f\}$ whose norms are $\leqslant A$ gives rise to a family of vectors $\{Tf\}$ which are equicontinuous and uniformly bounded. By the

Ascoli-Arzela theorem it follows that every sequence chosen from the collection $\{Tf\}$ contains a subsequence which converges *uniformly* (i.e., in the norm of $C([a, b])$. Thus, the operator T is completely continuous.

Theorem I: *Let B be any Banach space. Then*

(a) Any (finite) product of operators is completely continuous if at least one of the factors is completely continuous.

(b) Any (finite) linear combination of completely continuous operators is also completely continuous.

(c) The uniform limit of completely continuous operators is completely continuous. (Cf. Exercise 3.)

PROOF: (*a*) It obviously suffices to confine attention to a product of two operators, ST. If T is completely continuous and if $f_1, f_2, f_3, \ldots$ is any bounded sequence of vectors, there exists a subsequence of the sequence $Tf_1, Tf_2, Tf_3, \ldots$ which converges. Denoting this subsequence as $g_1, g_2, g_3, \ldots$ and its limit as g, we obtain $\|Sg - Sg_n\| = \|S(g - g_n)\| \leqslant \|S\| \cdot \|g - g_n\| \to 0$. Thus the sequence $Sg_1, Sg_2, Sg_3, \ldots$ is a convergent subsequence of the sequence $STf_1, STf_2, STf_3, \ldots$, and so ST is completely continuous.

On the other hand, if S is assumed to be completely continuous, we argue as follows. Given any bounded sequence of vectors $f_1, f_2, f_3, \ldots$, the sequence $Tf_1, Tf_2, Tf_3, \ldots$ is also bounded (since T is bounded), and so it is possible to extract from the sequence $S(Tf_1), S(Tf_2), S(Tf_3), \ldots$ a convergent subsequence. Therefore, ST is completely continuous.

(*b*) Since any scalar multiple of a completely continuous operator is obviously completely continuous, it suffices to prove that the sum of two completely continuous operators is also completely continuous, for finite induction then furnishes the desired result. Therefore, suppose that S and T are both completely continuous, and let $f_1, f_2, f_3, \ldots$ be any bounded sequence of vectors. Then we can select a subsequence $g_1, g_2, g_3, \ldots$ of this sequence such that the sequence $Tg_1, Tg_2, Tg_3, \ldots$ converges. From the sequence $g_1, g_2, g_3, \ldots$ we can, in turn, select a subsequence $h_1, h_2, h_3, \ldots$ such that the sequence $Sh_1, Sh_2, Sh_3, \ldots$ converges. Since the sequence $Th_1, Th_2, Th_3, \ldots$ certainly converges, it follows that the sequence $(S + T)h_1, (S + T)h_2, (S + T)h_3, \ldots$ also converges. Thus, the sequence $(S + T)f_1, (S + T)f_2, (S + T)f_3, \ldots$ contains a convergent subsequence, and so $S + T$ is completely continuous.

(*c*) Let the sequence $S_1, S_2, S_3, \ldots$ of completely continuous operators converge uniformly to the operator S and let $f_1, f_2, f_3, \ldots$ be any bounded sequence of vectors. From this latter sequence we extract a subsequence $f_{11}, f_{12}, f_{13}, \ldots$ such that $\lim_{n\to\infty} S_1 f_{1n}$ exists. From the sequence $f_{11}, f_{12}, f_{13}, \ldots$ we select a subsequence $f_{21}, f_{22}, f_{23}, \ldots$ such that $\lim_{n\to\infty} S_2 f_{2n}$ exists. Clearly, $\lim_{n\to\infty} S_1 f_{2n}$ also exists. Repeating this procedure, we obtain for each positive integer k a subsequence f_{k1}, f_{k2},

$f_{k3}, \ldots$ of the original sequence such that $\lim_{n\to\infty} S_j f_{kn}$ exists for $j = 1, 2, \ldots, k$. Now consider the diagonal sequence $f_{11}, f_{22}, f_{33}, \ldots$. This is clearly a subsequence of the original sequence $f_1, f_2, f_3, \ldots$, and, aside perhaps from a finite number (depending on k) of initial terms, it is a subsequence of the sequence $f_{k1}, f_{k2}, f_{k3}, \ldots$. Hence, $\lim_{n\to\infty} S_k f_{nn}$ exists for each index k. Given any $\epsilon > 0$, we first determine an index $\tilde{k}$ such that $\|S - S_{\tilde{k}}\| < \epsilon$, then an index N such that $\|S_{\tilde{k}}(f_{mm} - f_{nn})\| < \epsilon$ whenever m and n both exceed N. Under the restriction just imposed on m and n, we obtain

$$\begin{aligned}
\|Sf_{nn} - Sf_{mm}\| & \\
&= \|(Sf_{nn} - S_{\tilde{k}}f_{nn}) + (S_{\tilde{k}}f_{nn} - S_{\tilde{k}}f_{mm}) + (S_{\tilde{k}}f_{mm} - Sf_{mm})\| \\
&\leqslant \|(S - S_{\tilde{k}})f_{nn}\| + \|S_{\tilde{k}}f_{nn} - S_{\tilde{k}}f_{mm}\| + \|(S - S_{\tilde{k}})f_{mm}\| \\
&< \|S - S_{\tilde{k}}\| \cdot \|f_{nn}\| + \|S - S_{\tilde{k}}\| \cdot \|f_{mm}\| + \epsilon \\
&\leqslant 2\epsilon \sup \|f_n\| + \epsilon = \epsilon(1 + 2 \sup \|f_n\|).
\end{aligned}$$

Since ϵ can be chosen arbitrarily close to zero, it follows that the sequence $Sf_{11}, Sf_{22}, Sf_{33}, \ldots$ is Cauchy; since B is complete, the sequence is convergent, and so S is completely continuous.

Definition 2: *An operator defined on any normed linear space is said to be degenerate if its range (which is obviously a linear manifold) is finite-dimensional.*

Theorem 3: *Let T be a degenerate operator on the normed linear space N, and let the vectors $\{g_1, g_2, \ldots, g_n\}$ constitute a basis of the range of T. Then there exist bounded linear functionals $\{l_1, l_2, \ldots, l_n\}$ such that, for every vector f in N,*

$$Tf = l_1(f)g_1 + l_2(f)g_2 + \cdots + l_n(f)g_n.$$

PROOF: For any f, Tf must be expressible (in a unique manner) as a linear combination of the g's; the linearity of T clearly implies that the coefficients must be (scalar-valued) linear functions of f. It therefore remains only to show that the linear functionals $l_1, l_2, \ldots, l_n$ are bounded. By Theorem 2–4 of Chapter 4, we are assured of the existence of a positive constant δ such that $\|Tf\| \geqslant \delta \sum_{k=1}^{n} |l_k(f)|$, and so, for each index k, we obtain $|l_k(f)| \leqslant (1/\delta)\,\|Tf\|$. The boundedness of T furnishes the further inequality $|l_k(f)| \leqslant ((1/\delta)\,\|T\|)\,\|f\|$, and so the boundedness of l_k is established.

Corollary: *Any degenerate operator on any normed linear space is completely continuous.*

PROOF: Part (*b*) of Theorem 1 shows that it suffices to prove that, for any bounded linear functional l and any non-zero vector g, the equation $Tf = l(f)g$ defines a completely continuous operator T. Clearly, T is bounded and linear. Given any bounded sequence $f_1, f_2, f_3, \ldots$ we form the corresponding sequence of scalars $l(f_1), l(f_2), l(f_3), \ldots$. Each term of the sequence is bounded in absolute value by $\|l\| \cdot \sup \|f_n\|$, and by the Bolzano-Weierstrass theorem it is possible to extract a subsequence $\tilde{f}_1, \tilde{f}_2, \tilde{f}_3, \ldots$ of the original sequence of vectors such that $\lim_{n\to\infty} l(\tilde{f}_n)$ exists. Then clearly $T\tilde{f}_n \rightarrow \{\lim_{n\to\infty} l(\tilde{f}_n)\}g$, and so T is completely continuous.

Theorem 4: *Let T be a completely continuous operator on a Banach space and let μ be a non-zero eigenvalue of T. Then μ is of finite degeneracy. (The example $T = O$ shows that the non-vanishing of μ is an essential condition.)*

PROOF: Suppose that the manifold of all solutions of the equation $Tf = \mu f$ is of infinite dimension. By taking account of Theorem 2–7 of Chapter 4 we easily see that there would exist a sequence of unit eigenvectors $f_1, f_2, f_3, \ldots$ such that $\|f_i - f_j\| > \frac{1}{2}$ whenever $i \neq j$. Then, whenever $i \neq j$, the inequality $\|Tf_i - Tf_j\| = \|\mu(f_i - f_j)\| > \frac{1}{2}|\mu|$ would hold, and so the sequence $Tf_1, Tf_2, Tf_3, \ldots$ could not possibly contain a convergent sequence, but this would contradict the hypothesis that T is completely continuous.

Exercises

1. Prove the assertion made in the first sentence of Example (*c*).
2. Prove that the identity operator on any infinite-dimensional Banach space is not completely continuous.
3. Prove that part (*c*) of Theorem 1 becomes false if "uniform" is replaced by either "strong" or "weak."
4. Give an alternative discussion of Example (*d*) which is based on part (*c*) of Theorem 1, the corollary to Theorem 3, and the two-dimensional version of the Weierstrass approximation theorem (cf. Appendix *D*).
5. If T is an operator on a Hilbert space, show that T is completely continuous iff T^* is completely continuous.

§2. SPECTRAL ANALYSIS OF A HERMITIAN COMPLETELY CONTINUOUS OPERATOR

We saw in the previous chapter that every linear transformation on a finite-dimensional complex linear space has at least one eigenvalue.

Furthermore, under suitable hypotheses it was shown that it is possible to form a basis consisting of eigenvectors. However, we encounter difficulties immediately when we try to extend these results to infinite-dimensional spaces. Exercise 7–3 of Chapter 6 furnishes an example of a hermitian operator which possesses no eigenvalues, demonstrating that it is not possible to carry over to infinite-dimensional Hilbert spaces the elegant and simple description of the structure of Hermitian operators and the characterization of their eigenvalues and eigenvectors as solutions of certain extremal problems, which were obtained in the finite-dimensional case. However, we shall now see that the finite-dimensional theory of Hermitian operators does possess a simple generalization to infinite-dimensional Hilbert spaces when the additional condition of complete continuity is imposed.

Given on the infinite-dimensional Hilbert space H any hermitian operator T (not necessarily completely continuous), we can choose, exactly as in the finite-dimensional case, a sequence of unit vectors $g_1, g_2, g_3, \ldots$ such that $|(Tg_n, g_n)| \to \|T\|$, and since (Tg_n, g_n) is real we may assume further that $(Tg_n, g_n) \to \mu_1$, where $\mu_1 = \pm \|T\|$. Then

$$\begin{aligned}\|Tg_n - \mu_1 g_n\|^2 &= \|Tg_n\|^2 - 2\mu_1(Tg_n, g_n) + \mu_1^2 \|g_n\|^2 \\ &\leqslant \|T\|^2 + \mu_1^2 - 2\mu_1(Tg_n, g_n) = 2\mu_1^2 - 2\mu_1(Tg_n, g_n) \to 0.\end{aligned}$$

Hence, $Tg_n - \mu_1 g_n \to o$. We now impose on T the condition of complete continuity and select (if necessary) a subsequence of the g_n's, which we denote $h_1, h_2, h_3, \ldots$, such that the sequence $Th_1, Th_2, Th_3, \ldots$ converges. It then follows that the sequence $\mu_1 h_1, \mu_1 h_2, \mu_1 h_3, \ldots$ converges, and so (setting aside the trivial case $\mu_1 = 0$, corresponding to $T = O$) the sequence $h_1, h_2, h_3, \ldots$ must also converge; we denote the limit of this sequence as f_1. Then, referring back to the limiting relation $Tg_n - \mu_1 g_n \to o$, we conclude that $Tf_1 = \mu_1 f_1$. Since $\|f_1\| = 1$, we have shown that every operator T which is hermitian and completely continuous and not the zero operator possesses a non-zero eigenvalue, in particular $\pm \|T\|$.

The reader should have no difficulty in seeing that, exactly as in the finite-dimensional case, we may now restrict T to the orthogonal complement of the one-dimensional subspace spanned by f_1 and that if T does not reduce to the zero operator on the orthogonal complement, a second non-zero eigenvalue μ_2 and a corresponding (unit) eigenvector f_2 can be obtained; furthermore, $|\mu_2| \leqslant |\mu_1|$. By continuing this procedure as long as possible—that is, as long as T does not reduce to the zero operator on the orthogonal complement of the eigenvectors already obtained—we obtain either a finite or infinite sequence $\mu_1, \mu_2, \mu_3, \ldots$ of non-zero eigenvalues and a corresponding sequence $f_1, f_2, f_3, \ldots$ of eigenvectors.

If this procedure terminates in a finite number of steps (say after obtaining the eigenvalues $\mu_1, \mu_2, \ldots, \mu_N$ and eigenvectors $f_1, f_2, \ldots, f_N$), then every vector orthogonal to these eigenvectors is annihilated by T.

By choosing an orthonormal basis in the subspace of vectors orthogonal to the f_k's $(1 \leqslant k \leqslant N)$ and combining the two collections of orthonormal vectors, we obviously obtain an orthonormal basis of the entire Hilbert space. Furthermore, every vector f can be expressed in the form

$$f = \sum_{k=1}^{N} (f, f_k) f_k + \tilde{f},$$

where $T\tilde{f} = o$, and so

$$Tf = \sum_{k=1}^{N} (f, f_k) Tf_k = \sum_{k=1}^{N} \mu_k (f, f_k) f_k. \tag{2–1}$$

On the other hand, if an infinite sequence $\mu_1, \mu_2, \mu_3, \ldots$ of non-zero eigenvalues is obtained, it may still occur that the corresponding eigenvectors $f_1, f_2, f_3, \ldots$ do not span the entire Hilbert space. (Cf. Exercise 4.) However, whether or not the eigenvectors $f_1, f_2, f_3, \ldots$ span the space, it is easily seen that the obvious generalization of (2–1) holds:

$$Tf = \sum_{k=1}^{\infty} \mu_k (f, f_k) f_k. \tag{2-2}$$

Next, we point out that, when an infinite sequence of non-zero eigenvalues is obtained, the limiting relation $\lim_{k\to\infty} \mu_k = 0$ must hold. The proof is very simple. Since the procedure employed guarantees that the chain of inequalities $|\mu_1| \geqslant |\mu_2| \geqslant |\mu_3| \geqslant \cdots$ must hold, the quantity $\lim_{k\to\infty} |\mu_k|$ must exist and be non-negative. If the limit were a positive number, say α, then the following inequality would be valid for any two distinct indices j and k (since $(f_j, f_k) = 0$):

$$\|Tf_j - Tf_k\|^2 = \|\mu_j f_j - \mu_k f_k\|^2 = \mu_j^2 + \mu_k^2 \geqslant 2\alpha^2.$$

Thus, the sequence $Tf_1, Tf_2, Tf_3, \ldots$ could not contain a convergent subsequence, contradicting the hypothesis of complete continuity.

We now turn to the task of generalizing equations (5–7) and (5–8) of Chapter 7. For any real number λ we define $\mathfrak{M}_\lambda$ as the subspace spanned by the eigenvectors associated with the eigenvalues of T which are $\leqslant \lambda$. It is readily seen that $\mathfrak{M}_\lambda$ is finite-dimensional if λ is negative and that $\mathfrak{M}_\lambda^\perp$ is finite-dimensional if λ is positive; in particular, $\mathfrak{M}_\lambda = \{o\}$ if $\lambda < -\|T\|$ and $\mathfrak{M}_\lambda = H$ if $\lambda \geqslant \|T\|$. We then define P_λ as the projection on $\mathfrak{M}_\lambda$. Exactly as in the finite-dimensional case, we refer to $\{P_\lambda\}$ as the resolution of the identity associated with the operator T, and with very little effort (cf. Exercise 7) we can justify the following formula for (Tf, g) as a Stieltjes integral, for any vectors f and g in H:

$$(Tf, g) = \int_{-\infty}^{\infty} \lambda \, d(P_\lambda f, g). \tag{2–3}$$

While (2–3) is identical in appearance with (5–7) of Chapter 7, it should be emphasized that the integral appearing in the latter equation is, in reality, a finite sum, while the integral in (2–3) is, except in trivial cases (namely, when the operator T is degenerate) an infinite series. Now, we are justified in rewriting (2–3) in the following form, identical in appearance with (5–8) of Chapter 7:

$$T = \int_{-\infty}^{\infty} \lambda \, dP_\lambda. \tag{2–4}$$

The equation (2–4) (which is actually nothing but a reformulation of (2–2)) is known as the *spectral representation* of the operator T, and the fact that corresponding to every hermitian completely continuous operator T there exists a resolution of the identity $\{P_\lambda\}$ such that the representation (2–4) holds is known as the spectral theorem (for such operators).

We conclude this section by indicating, very briefly and imprecisely, how this restricted version of the spectral theorem can be extended. By a much more delicate argument than that which we have employed, it can be shown that, even without the hypothesis of complete continuity, there may be associated with any hermitian operator T a resolution of the identity $\{P_\lambda\}$ such that (2–3) and (2–4) hold. However, in the completely continuous case the *points of increase* of P_λ (the numbers λ such that $\mathfrak{M}_\mu \subset \mathfrak{M}_\nu$ whenever $\mu < \lambda < \nu$) are isolated, with the possible exception of zero, while in the general hermitian case the set formed by these points (which, as might be expected, coincides with the spectrum) may have a much more complicated structure. Exercise 8 provides a simple, but not entirely trivial, example of the spectral representation of an operator which is hermitian but not completely continuous.

Once the spectral theorem is developed for hermitian operators, it is comparatively easy to develop a spectral representation for a normal operator as an integral over the complex plane.

An excellent presentation of the Spectral Theorem will be found in [Riesz and Sz.-Nagy, Chapter 7].

Exercises

1. Let T be an operator on a Banach space. The scalar λ is said to be an *approximate eigenvalue* of the operator T if λ is not an eigenvalue of T but if, for every positive number ϵ, there exists a non-zero vector f such that $\|Tf - \lambda f\| < \epsilon \|f\|$. Show that, for the operator T appearing in Exercise 7–3 of Chapter 6, every number in the spectrum is an approximate eigenvalue.

2. Let T be the *shift-operator* on the Hilbert space l^2, defined as follows: For every vector $(a_1, a_2, a_3, \ldots)$ in l^2,

$$T(a_1, a_2, a_3, \ldots) = (0, a_1, a_2, a_3, \ldots).$$

Prove that the number zero belongs to the spectrum of T but that it is neither an eigenvalue nor an approximate eigenvalue of T.

3. In contrast to the preceding exercise, show that every number in the spectrum of a hermitian operator is either an eigenvalue or an approximate eigenvalue. (Note that the shift-operator is not hermitian; what is its adjoint?)

4. In the Hilbert space l^2, give examples of hermitian completely continuous operators T_1, T_2, T_3, T_4 satisfying the following conditions:

(*a*) T_1 has only a finite number of non-zero eigenvalues.

(*b*) T_2 has an infinite number of non-zero eigenvalues, and the corresponding eigenvectors span l^2.

(*c*) T_3 has an infinite number of non-zero eigenvalues, and the orthogonal complement of the subspace spanned by the corresponding eigenvectors is finite-dimensional.

(*d*) T_4 has an infinite number of non-zero eigenvalues and the orthogonal complement of the subspace spanned by the corresponding eigenvectors is infinite-dimensional.

5. Show that the procedure described in the text furnishes all the non-zero eigenvalues with the correct multiplicity.

6. Prove that the spectrum of a hermitian completely continuous operator (on an infinite-dimensional Hilbert space) must contain the number zero. If the latter is an isolated point of the spectrum, it is an eigenvalue, but if it is not isolated it may be either an eigenvalue or an approximate eigenvalue.

7. Justify equation (2–3).

8. Let T be the operator appearing in Exercise 7–3 of Chapter 6 and let $p_\lambda(x)$, for each real number λ and each number x in $[0, 1]$, be defined as follows: $p_\lambda(x) \equiv 0$ if $\lambda < 0$, $p_\lambda(x) \equiv 1$ if $\lambda \geqslant 1$, and p_λ is the characteristic function of the interval $[0, \lambda]$ if $0 \leqslant \lambda \leqslant 1$. Let P_λ denote the operator (on $L^2([0, 1])$) consisting of multiplication by p_λ. Show that $\{P_\lambda\}$ is a resolution of the identity on $L^2([0, 1])$ and that (2–3) (and hence (2–4)) holds. (Note that the points of increase of $\{P_\lambda\}$ constitute the entire interval $[0, 1]$, which is precisely the spectrum of T.)

§3. THE FREDHOLM ALTERNATIVE

As in the preceding section, let T be a hermitian completely continuous operator, let λ be a given scalar, and let g be a given vector. Consider the problem

$$f = g + \lambda T f. \tag{3–1}$$

(We shall see in the following section a particularly important example of such a problem.) Taking account of (2–1) and (2–2), we see that (3–1) can be rewritten as follows:

$$f = g + \sum_k \lambda\mu_k(f, f_k)f_k, \tag{3–2}$$

the summation being performed over all the non-zero eigenvalues of T. If we now replace f on *both* sides of (3–2) by the *right* side of this equation, we obtain (by employing due caution with the indices)

$$g + \sum_k \lambda\mu_k(f, f_k)f_k = g + \sum_k \sum_j \lambda^2\mu_j\mu_k(f, f_j)(f_j, f_k)f_k + \sum_k \lambda\mu_k(g, f_k)f_k. \tag{3–3}$$

This simplifies, because of the orthonormality of the eigenvectors, to

$$\sum_k \lambda\mu_k(f, f_k)f_k = \sum_k \{\lambda^2\mu_k^2(f, f_k) + \lambda\mu_k(g, f_k)\}f_k. \tag{3–4}$$

Since the eigenvectors are linearly independent, we obtain

$$\lambda\mu_k(1 - \lambda\mu_k)(f, f_k) = \lambda\mu_k(g, f_k). \tag{3–5}$$

Now, none of the μ_k's vanishes, and we may assume that $\lambda \neq 0$, for otherwise (3–1) becomes utterly trivial. Hence, we may cancel out the factor $\lambda\mu_k$ from both sides of (3–5), obtaining

$$(1 - \lambda\mu_k)(f, f_k) = (g, f_k). \tag{3–6}$$

Now, there are two possibilities:

(*a*) $1 - \lambda\mu_k$ never vanishes—i.e., λ is not equal to the reciprocal of any eigenvalue of T. In this case, (f, f_k) must equal, for each index k, the quantity $(g, f_k)/(1 - \lambda\mu_k)$, and so the only conceivable solution of (3–1) is obtained by inserting into (3–2) the expression just obtained for (f, f_k). We thus obtain for the sole possible solution the formula

$$f = g + \sum_k \frac{\lambda\mu_k(g, f_k)f_k}{1 - \lambda\mu_k}. \tag{3–7}$$

(*b*) $1 - \lambda\mu_k$ vanishes for one or more indices k; as we have seen previously, this can occur for only a *finite* number of indices. In this case, we see from (3–6) that there can be no solution of (3–1) unless (g, f_k) vanishes for all the eigenvectors f_k associated with the eigenvalue $1/\lambda$; if this condition is satisfied, then (3–6) imposes *no* restriction on the quantities (f, f_k), and so we are led, by again referring to (3–2), to the following formula for the most general possible solution of (3–1):

$$f = g + \sum_{\mu_k \neq 1/\lambda} \frac{\lambda\mu_k(g, f_k)f_k}{1 - \lambda\mu_k} + \sum_{\mu_k = 1/\lambda} \gamma_k f_k, \tag{3–8}$$

where the scalars γ_k are completely arbitrary.

It still remains to show that the formulas (3–7) and (3–8) are meaningful. If T possesses only a finite number of non-zero eigenvalues, no problem of convergence arises, and by direct substitution into (3–1) we see that (3–7) or (3–8) does indeed provide a solution of (3–1). On the other hand, if T possesses infinitely many non-zero eigenvalues, the fact that $\lim_{k\to\infty} \mu_k = 0$, which we proved in §2, enables us to assert that the inequality $|\lambda\mu_k| < \frac{1}{2}$ holds with only a finite number of exceptional values of k (if any). Thus, setting aside these exceptional values of k, we conclude that the coefficient of f_k appearing in (3–7) or in the infinite summation in (3–8) is dominated in absolute value by $\frac{1}{2}|(g, f_k)|/(1 - \frac{1}{2})$, or $|(g, f_k)|$. Since, by Bessel's inequality, $\sum_k |(g, f_k)|^2$ is convergent, the aforementioned infinite sums do converge (in the norm of the Hilbert space under consideration), and hence the right sides of (3–7) and (3–8) are indeed meaningful. Then, as in the case when T possesses only finitely many non-zero eigenvalues, we easily confirm that (3–7) or (3–8) does provide a solution of (3–1). Thus, we have established the following result.

Theorem 1: *If T is a hermitian completely continuous operator* (*on an infinite-dimensional Hilbert space*) *and if λ is not equal to the reciprocal of any of the eigenvalues of T, then for each vector g the equation* (3–1) *possesses one and only one solution, given by* (3–7). *If $1/\lambda$ is an eigenvalue of T, then the equation* (3–1) *is solvable iff g is orthogonal to the subspace $\mathfrak{M}$* (*necessarily finite-dimensional*) *spanned by the eigenvectors associated with the eigenvalue $1/\lambda$* (*in other words, iff g is orthogonal to all solutions of the equation $Th = (1/\lambda)h$*). *If g satisfies this condition,* (3–1) *possesses infinitely many solutions; exactly one of these is orthogonal to $\mathfrak{M}$, and the general solution is obtained by adding to this particular solution an arbitrary vector contained in $\mathfrak{M}$.*

If the hypothesis that T is hermitian is omitted, the preceding argument becomes completely inapplicable; nevertheless, the very remarkable fact holds that Theorem 1 can be recast into a form which is valid without this hypothesis. We shall not present the proof, but shall merely remark that the proof is obtained by beginning with the observation that the theorem is known by linear algebra to be true in the finite-dimensional case and then reducing the infinite-dimensional theorem to the finite-dimensional case by approximating the operator T with sufficient accuracy by a degenerate operator.

Theorem 2 (Fredholm Alternative in Hilbert Space): *Let T be completely continuous. The equation* (3–1) *is uniquely solvable for every vector g iff the equation $f = \lambda Tf$ (corresponding to $g = o$) possesses only the trivial solution $f = o$. This condition, in turn, is equivalent to the condition that the adjoint homogeneous equation $f = \bar{\lambda}T^*f$ shall possess only the trivial solution.*

If the equation $f = \lambda Tf$ possesses non-trivial solutions, so does the equation $f = \bar{\lambda}T^*f$, and the solutions of these two equations form linear manifolds, $\mathfrak{M}$ and $\mathfrak{M}^*$, of the same (finite) dimension. In this case the equation (3–1) *is solvable iff g is orthogonal to $\mathfrak{M}^*$; when this condition is satisfied,* (3–1) *possesses infinitely many solutions, exactly one of which is orthogonal to $\mathfrak{M}$, and the general solution is obtained by adding to this particular solution an arbitrary member of $\mathfrak{M}$.*

Finally, we discuss briefly the remarkable fact that a major part of this theorem can be carried over to Banach spaces.

Theorem 3 (Fredholm Alternative in Banach Space): *Let T be completely continuous. The equation* (3–1) *is uniquely solvable for every vector g iff the equation $f = \lambda Tf$ (corresponding to $g = o$) possesses only the trivial solution $f = o$. If the latter equation possesses a non-trivial solution, then there exist vectors g for which* (3–1) *is not solvable.*

(Note that the first two sentences of this theorem and of Theorem 2 are identical.)

As might be expected, the proof of this theorem is somewhat more subtle than the proof of the corresponding portion of Theorem 2, since much of the structural simplicity possessed by Hilbert spaces (centering around the concept of orthogonality) is lost is making the transition to Banach spaces. Nevertheless, it is even possible to extend Theorem 3 so as to provide analogues of the remaining portions of Theorem 2. As might be expected, the analogues will involve the adjoint operator T^*, which operates on the dual space rather than on the original Banach space.

Exercises

1. Use formula (3–7) to show that when λ is not the reciprocal of any eigenvalue of T, the transformation $(I - \lambda T)^{-1}$ is bounded. (Of course, this is also guaranteed by the second corollary to Theorem 3–1 of Chapter 6.)

2. Show by means of a simple example that none of the theorems stated in this section remains correct if the condition of complete continuity is omitted.

§4. SURVEY OF THE FREDHOLM THEORY OF INTEGRAL EQUATIONS

Fredholm's own discovery, which was made early in this century, antedates the development of the theory of Hilbert and Banach spaces. The major part of his theorem, as he formulated it, goes as follows: If $[a, b]$ is a compact interval, if g is a (real-valued or complex-valued) continuous function on $[a, b]$, if K is a (real-valued or complex-valued) continuous function on the square $[a, b] \times [a, b]$, and if the only continuous solution of the equation

$$f(x) = \int_a^b K(x, y)f(y)\,dy \tag{4–1}$$

is the trivial one, $f(x) \equiv 0$, then the equation

$$f(x) = g(x) + \int_a^b K(x, y)f(y)\,dy \tag{4–2}$$

possesses a unique continuous solution.

By referring to Example (d) of §1, we immediately see that Fredholm's theorem is contained as a particular case of Theorem 3–3. Needless to say, Fredholm's remarkable discovery served as a major stimulus to the line of abstract development which culminated in the proof of Theorem 3–3. It should, however, be pointed out that the proof of Theorem 3–3 does not run parallel to Fredholm's own argument. Briefly, Fredholm partitioned the interval $[a, b]$ into subintervals of equal length and approximated the integral appearing in (4–2) by a sum; he was thus led to approximate (4–2) by a finite system of linear algebraic equations, and then with remarkable skill he carried out a passage to the limit in which the number of subintervals increased without bound. In particular, he dealt with determinants (via Cramer's rule) of increasing size. The proof of the abstract theorem makes no reference to determinants, or even to the finite-dimensional case; the theory of (finite) systems of algebraic equations drops out as a particular case of the general theorem, as does Fredholm's own theorem.

We now discuss briefly and heuristically how the Hilbert-space formulation of the Fredholm alternative applies to integral equations. A rigorous discussion must be based on the theory of Lebesgue integration in the plane, which we have only sketched briefly in Chapter 2. In particular, as will be seen in the following discussion, it is necessary to appeal frequently to the theorem of Fubini.

Let the function K be measurable and quadratically integrable on the square $(a, b) \times (a, b)$, which we denote for brevity by D. (The interval

(a, b) may be infinite in either or both directions, in contrast to the Banach-space formulation of the Fredholm theory.) Then for any function f belonging to $L^2((a, b))$ the integral $\int_a^b K(x, y)f(y)\,dy$ exists for almost all x in (a, b) and the function thus defined, which for obvious reasons is denoted $\mathbf{K}f$, is also a member of $L^2((a, b))$. Furthermore, by using the Schwarz inequality in combination with Fubini's theorem, we can show that $\|\mathbf{K}f\|^2 \leqslant \{\iint_D |K|^2\} \|f\|^2$. Thus, the kernel K gives rise to a *bounded* linear transformation $\mathbf{K}$ (i.e., an operator) on $L^2((a, b))$. By suitably approximating the kernel K with polynomials, which are easily seen to give rise to *degenerate* operators, and employing part (*c*) of Theorem 1–1 in combination with the corollary to Theorem 1–3, we then see that the operator $\mathbf{K}$ is completely continuous, and so Theorem 3–2 may be applied to the integral equation

$$f(x) = g(x) + \int_a^b K(x, y)f(y)\,dy, \tag{4–3}$$

where the given function g and the unknown function f are to belong to $L^2((a, b))$.

Next, it is almost obvious (cf. Exercise 4–3) that the *operator* $\mathbf{K}$ is hermitian if the *kernel* K is hermitian—that is, if $K(x, y) = \bar{K}(y, x)$ for almost all points (x, y) in the square D. (The converse is also true.) Now Theorem 3–1 and all of §2 are applicable. In particular, if the operator $\mathbf{K}$ corresponding to the kernel has only a finite number of non-zero eigenvalues $\mu_1, \mu_2, \ldots, \mu_N$ (each appearing as frequently as its multiplicity), with corresponding orthonormal eigenvectors (which are now termed *eigenfunctions*) $f_1, f_2, \ldots, f_N$, we readily see from the developments of §2 that the kernel (also obviously hermitian)

$$\tilde{K}(x, y) = K(x, y) - \sum_{k=1}^{N} \mu_k f_k(x)\overline{f_k(y)} \tag{4–4}$$

must give rise to the zero operator. Now, as might be expected (although we do not prove it here), this implies that $\tilde{K}(x, y) = 0$ almost everywhere in D, and so from (4–4) we obtain the remarkable *bilinear expansion*

$$K(x, y) = \sum_{k=1}^{N} \mu_k f_k(x)\overline{f_k(y)} \quad \text{a.e.} \tag{4–5}$$

(Cf. Exercise 4–4.) On the other hand, if K possesses infinitely many non-zero eigenvalues $\mu_1, \mu_2, \ldots$ (arranged in descending order of absolute value, as in §2), it is clear that, for any positive integer n, the kernel

$$K(x, y; n) = K(x, y) - \sum_{k=1}^{n} \mu_k f_k(x)\overline{f_k(y)} \tag{4–6}$$

gives rise to an operator $\mathbf{K}_n$ whose norm is precisely $|\mu_{n+1}|$, so that $\|\mathbf{K}_n\| \to 0$. This property of the *operators* $\mathbf{K}_n$ suggests very strongly that a corresponding property must hold for the kernels $K(x, y; n)$—i.e., that as n increases without bound the limiting relation

$$\iint_D |K(x, y; n)|^2 = \iint_D \left| K(x, y) - \sum_{k=1}^{n} \mu_k f_k(x) \overline{f_k(y)} \right|^2 \to 0 \qquad (4\text{–}7)$$

must hold. This assertion is indeed true, so that we may write the following generalization of (4–5):

$$K(x, y) = \sum_{k=1}^{\infty} \mu_k f_k(x) \overline{f_k(y)}. \qquad (4\text{–}8)$$

However, it must be emphasized that (4–8) does not assert that the series on the right converges *pointwise* to the left side, but only that the partial sums of the right side converge to the left side in the norm of $L^2(D)$. However, as might be expected, there are many important particular kernels for which (4–8) holds in the sense of pointwise convergence as well as in the sense of convergence in norm.

We conclude our discussion of the expansion theory of hermitian kernels at this point, but we strongly advise that the reader pay careful attention to Exercises 5 through 9, in which some of the ideas just presented are pushed further.

Exercises

1. Let K be continuous on the compact square $[a, b] \times [a, b]$ and suppose that $|\lambda| \cdot (b - a) \cdot \max |K(x, y)| < 1$. Prove that the integral equation (4–2) possesses a unique continuous solution f for each continuous function g.

2. Let K be defined and continuous on the closed triangular region bounded by the x-axis, the line $x = y$, and the line $x = a$, where a is a positive constant. Show that the integral $\int_0^x K(x, y) f(y)\, dy$ defines a completely continuous operator $\mathbf{K}$ on $C([0, a])$ and that the integral equation

$$f(x) = g(x) + \lambda \int_0^x K(x, y) f(y)\, dy$$

possesses a unique continuous solution f in $C([0, a])$ for each g in $C([0, a])$ and for any value of λ. (Hint: Prove that for some sufficiently large integer n the norm of the operator $\mathbf{K}^n$ is less than unity.) Equations of this type are known as *Volterra equations*.

3. Prove that a hermitian (quadratically integrable) kernel gives rise to a hermitian operator on $L^2((a, b))$.

4. Let $K(x, y) = \cos(x + y)$ on the square $(0, \pi) \times (0, \pi)$. Show that the corresponding operator $\mathbf{K}$ has only two non-zero eigenvalues (each non-degenerate). Determine these eigenvalues and the corresponding unit eigenvectors (eigenfunctions) and confirm that (4–5) holds in this case.

Note: The remaining exercises are to be solved in the sense of resorting to formal manipulations whose rigorous justification would necessitate an adequate command of the theory of Lebesgue integration in the plane.

5. Show that if $K(x, y)$ and $L(x, y)$ belong to $L^2(D)$, then the integral $\int_a^b K(x, t)L(t, y)\, dt$ defines a kernel $M(x, y)$ which also belongs to $L^2(D)$, and that the corresponding operators $\mathbf{K}$, $\mathbf{L}$, $\mathbf{M}$ satisfy the relation $\mathbf{M} = \mathbf{KL}$. Furthermore, show by means of a specific example that hermitian kernels $K(x, y)$ and $L(x, y)$ may furnish a non-hermitian kernel $M(x, y)$.

6. If $K(x, y)$ is hermitian and belongs to $L^2(D)$, show that, in analogy with (4–8) (or (4–5)) the relation

$$K_{(2)}(x, y) = \sum_{k=1}^{N \text{ or } \infty} \mu_k^2 f_k(x)\overline{f_k(y)} \tag{4–9}$$

holds, where $K_{(2)}(x, y) = \int_a^b K(x, t)K(t, y)\, dt$. More generally, show that, with an obvious appropriate definition of $K_{(n)}(x, y)$, the relation

$$K_{(n)}(x, y) = \sum_{k=1}^{N \text{ or } \infty} \mu_k^n f_k(x)\overline{f_k(y)} \tag{4–10}$$

holds for every positive integer n.

7. Employing the notation introduced in the preceding exercise, show that

$$\iint_D |K_{(n)}(x, y)|^2 = \sum_{k=1}^{N \text{ or } \infty} \mu_k^{2n}. \tag{4–11}$$

8. Let $K(x, y) = \min\{x, y\} \cdot [1 - \max\{x, y\}]$ on the square $(0, 1) \times (0, 1)$. Show that the eigenvalues of the corresponding operator are all simple and positive, being given by the formula

$$\mu_k = \frac{1}{k^2\pi^2} \qquad (k = 1, 2, 3, \ldots), \tag{4–12}$$

while the corresponding unit eigenvectors are given by the functions

$$f_k(x) = \sqrt{2}\sin k\pi x. \tag{4–13}$$

9. Work out the numerical value of the left side of (4–11) for the cases $n = 1$ and $n = 2$, where $K(x, y)$ is the kernel appearing in Exercise 8. Derive from your calculations the identities

$$\sum_{k=1}^{\infty}\frac{1}{k^4} = \frac{\pi^4}{90}, \qquad \sum_{k=1}^{\infty}\frac{1}{k^8} = \frac{\pi^8}{9450}. \tag{4–14}$$

§5. ESTIMATION OF THE LARGEST EIGENVALUE

From equations (2–1) and (2–2) it is almost immediately evident that the discussion of Kellogg's method which was presented in the preceding chapter carries over with no essential changes to the infinite-dimensional case, except that we must now impose on the hermitian operator T the condition of complete continuity. We illustrate this procedure with the operator determined by the kernel $K(x, y)$ defined in Exercise 4–8. According to Exercise 1, the norm of the operator $\mathbf{K}$ determined by this kernel is certainly an eigenvalue, and the (essentially unique) eigenfunction f_1 corresponding to this eigenvalue may be taken to be everywhere real and non-negative, so that any non-zero constant function (which we may take to be $f \equiv 1$) is certainly not orthogonal to f_1; since $-\ \|\mathbf{K}\|$ is certainly not an eigenvalue (again according to Exercise 1), we may be assured that the sequence of numbers $(\mathbf{K}^{n+1}f, f)/(\mathbf{K}^n f, f)$ converges to $\|\mathbf{K}\|$, while the positive character of the operator $\mathbf{K}$ (cf. Exercise 4–8) guarantees the *monotone* convergence of the aforementioned sequence to $\|\mathbf{K}\|$.

Now, by elementary calculations (which the reader should carry out for his own benefit) we obtain, with the aforementioned choice of the function f, the following results:

$$\mathbf{K}f(x) = \tfrac{1}{2}(x - x^2), \quad \mathbf{K}^2 f(x) = \tfrac{1}{24}(x - 2x^3 + x^4), \tag{5–1}$$
$$\mathbf{K}^3 f(x) = \tfrac{1}{720}(3x - 5x^3 + 3x^5 - x^6).$$

Next, we easily obtain

$$(f, f) = 1, \quad (\mathbf{K}f, f) = \tfrac{1}{12}, \quad (\mathbf{K}^2 f, f) = \tfrac{1}{120}, \tag{5–2}$$
$$(\mathbf{K}^3 f, f) = \tfrac{17}{20160}.$$

From these results we obtain, in turn,

$$\frac{(\mathbf{K}f, f)}{(f, f)} = \frac{1}{12}, \quad \frac{(\mathbf{K}^2 f, f)}{(\mathbf{K}f, f)} = \frac{1}{10}, \quad \frac{(\mathbf{K}^3 f, f)}{(\mathbf{K}^2 f, f)} = \frac{17}{168}, \tag{5–3}$$

and we observe that these ratios are increasing. It is interesting to note that $(1/\pi^2)/(\frac{17}{168}) = 1.00129\ldots$, so that in only three stages the Kellogg procedure has furnished an estimate for the largest eigenvalue which is accurate within almost 0.1 percent, despite the fact that the constant function f is a rather poor approximation for the eigenfunction corresponding to the largest eigenvalue.

Continuing our calculations a little further, we obtain

$$\frac{\mathbf{K}f(x)}{\|\mathbf{K}f\|} = \sqrt{30}(x - x^2), \quad \frac{\mathbf{K}^2f(x)}{\|\mathbf{K}^2f\|} = \sqrt{\frac{630}{31}}\,(x - 2x^3 + x^4),$$
$$\frac{\mathbf{K}^3f(x)}{\|\mathbf{K}^3f\|} = \sqrt{\frac{12012}{5461}}\,(3x - 5x^3 + 3x^5 - x^6), \tag{5–4}$$

and also, recalling that the leading eigenfunction f_1 is given by $f_1(x) = \sqrt{2}\sin \pi x$, we find, by a tedious but elementary computation, that

$$\left\| \frac{\mathbf{K}^3f}{\|\mathbf{K}^3f\|} - f_1 \right\|^2 = 2 - \frac{5760}{\pi^7}\left(\frac{6006}{5461}\right)^{1/2} \approx 0.20924 \cdot 10^{-6}, \tag{5–5}$$

or

$$\left\| \frac{\mathbf{K}^3f}{\|\mathbf{K}^3f\|} - f_1 \right\| \approx 0.4574 \cdot 10^{-3}. \tag{5–6}$$

Now, instead of choosing a trial function at random, as we did previously, we turn to the problem of making an optimum choice of the trial function within a specified class of candidates. From elementary considerations of symmetry, it is clear that the (essentially unique) eigenfunction of the kernel under discussion here must be symmetric with respect to the point $\frac{1}{2}$; i.e., $f_1(\frac{1}{2} + x) = f_1(\frac{1}{2} - x)$. Let us therefore consider all quadratic polynomials which possess this symmetry property. Since the unique critical value of such a function must occur at $\frac{1}{2}$, we see that it must be of the form $a + bx - bx^2$. An elementary calculation furnishes the result

$$\frac{(\mathbf{K}f, f)}{(f, f)} = \frac{420a^2 + 17b^2 + 168ab}{168(30a^2 + b^2 + 10ab)}, \tag{5–7}$$

or

$$\frac{(\mathbf{K}f, f)}{(f, f)} = \frac{420 + 168c + 17c^2}{168(30 + 10c + c^2)}, \qquad c = \frac{b}{a}. \tag{5–8}$$

A simple calculation shows that this fraction* is maximized by choosing

* Fractions of the form $(\mathbf{K}f, f)/(f, f)$ are frequently termed *Rayleigh quotients*, particularly when they are employed in estimating eigenvalues.

for c the value $-45 - \sqrt{1605}$, and that with this choice of c equation (5–8) assumes the form

$$\frac{(\mathbf{K}f, f)}{(f, f)} = \frac{4815 + 107\sqrt{1605}}{89880} = 0.1012648 \cdots. \tag{5–9}$$

Taking account of the characterization of λ_1 as the maximum of the fraction $(\mathbf{K}f, f)/(f, f)$ for all possible choices of f in $L^2((0, 1))$ (subject, of course, to the restriction $\|f\| > 0$), we conclude that

$$\lambda_1 \geqslant 0.1012648 \cdots. \tag{5–10}$$

It is interesting to note that $\lambda_1 = 1/\pi^2 = 0.101321 \ldots$, so that consideration of a very simple class of trial functions has led us to an estimate for the largest eigenvalue which is in error by only 0.05 percent, a result better than that obtained with three stages of the Kellogg procedure (beginning, however, with a less judicious choice of trial function).

The method of employing a family of trial functions containing one or more parameters and making an optimum choice of the parameter(s) is known as the Rayleigh-Ritz procedure, and it can often be used very effectively in estimating the highest eigenvalue (and, with suitable modifications, lower eigenvalues) of hermitian kernels.

Finally, we remark that, once the Rayleigh-Ritz procedure has furnished a satisfactory trial function, the latter may be effectively employed in the Kellogg procedure.

Exercises

1. Let $K(x, y)$ be real-valued, quadratically integrable, and symmetric (i.e., $K(x, y) = K(y, x)$ a.e.) on a square $(a, b) \times (a, b)$, and suppose furthermore that $K(x, y)$ is positive everywhere (or almost everywhere) on the square. Show that $\|\mathbf{K}\|$ is an eigenvalue of the corresponding operator $\mathbf{K}$, that this eigenvalue is simple, that associated with this eigenvalue there is a non-negative eigenfunction, and that $-\|\mathbf{K}\|$ is not an eigenvalue. (Hint: Note the strong resemblance to Exercise 3–2 of Chapter 7.)

2. Let $f(x) = \gamma\{1 - (45 + \sqrt{1605})(x - x^2)\}$, where the real number γ is to be chosen so that $\|f(x)\| = 1$. (Note that the sign of γ is left ambiguous.) Compute $\|\sqrt{2} \sin \pi x - f(x)\|$, where the sign of γ is chosen so as to make this norm close to zero. (Clearly, the opposite choice of γ will make the norm close to 2.)

APPENDICES

A. PARTIALLY ORDERED SETS AND ZORN'S LEMMA

Let A be any non-empty set and let P be any non-empty subset of $A \times A$ (i.e., P is a relation on A). If the elements x and y of A are such that the ordered pair (x, y) belongs to P, we shall write xPy. (In particular, P may coincide with $A \times A$, in which case the statement xPy is true for every choice of x and y.) The elements x and y (not necessarily distinct) are said to be *comparable* if at least one of the statements xPy, yPx is true; otherwise they are said to be incomparable. The relation P is said to be a *partial ordering* (of A) if the following three conditions are satisfied:

(*a*) xPx for every x (*axiom of reflexivity*);

(*b*) If xPy and yPx, then $x = y$ (*axiom of symmetry*);

(*c*) If xPy and yPz, then xPz (*axiom of transitivity*).

The simplest and most important example of a partial ordering is the relation $\leqslant$ on the real number system R. (For this reason, it is customary to use the symbol $\leqslant$ for any partial ordering, but we prefer to employ the non-committal symbol P.) Note carefully that in this case any two elements are comparable. On the other hand, if we still consider R but interpret xPy to mean that $y - x$ is a non-negative rational number, then P is indeed a partial ordering, but the numbers 1 and $\sqrt{2}$ are incomparable.

The set A, together with the partial ordering P, is termed a partially ordered set, or *poset*. (Strictly speaking one should describe a poset as a pair of objects $\{A, P\}$, but, as in many analogous situations, we speak simply of the poset A, suppressing reference to the partial ordering P once it has been defined.) If every two elements of a poset A are comparable, then A is said to be *totally ordered*, or *linearly ordered*.

A non-empty subset B of the poset A is said to be a *chain* if every two members of B are comparable (i.e., if the restriction of P to $B \times B$ constitutes a total ordering of B). For example, if we refer to the second example of a partial ordering of the real number system, we easily see that the set of all rational numbers constitutes a chain; in fact, any non-empty subset of the rational numbers constitutes a chain.

Next, let B be any non-empty subset (not necessarily a chain) of the poset A and suppose that the element x of A (not necessarily belonging to B) has the property that yPx for every element y belonging to B. Then x is said to be an *upper bound* of B. (Note that an upper bound need not exist; for example, if B denotes the set of all real negative numbers, then zero is an upper bound of B under the first, or usual, partial ordering of R, while the same set has no upper bound under the second ordering, since no number could be comparable with both the rational number -1 and the irrational number $-\sqrt{2}$.) Observe that we shall say "B has an upper bound" even though B may not *contain* an upper bound.

The last concept to be introduced is that of a maximal element. Suppose that the element x of the poset A has the property that there is no element y of A other than x such that xPy (roughly speaking, no element of A is strictly larger than x). Then x is termed a *maximal element* of A. It should be emphasized that a maximal element x does not necessarily have the property that yPx for every element y of A. For example, let A consist of the four elements a, b, c, d. Then the collection P of ordered pairs $\{(a, a), (a, b), (b, b), (c, c), (c, d), (d, d)\}$ constitutes a partial ordering; both b and d are maximal elements, yet the relation aPd does not hold, so that it is not true that d is strictly larger than all other members of A.

We are now prepared to state Zorn's Lemma: *If a poset A has the property that every chain in A has an upper bound, then A contains at least one maximal element.*

The following list of exercises should help the reader develop a firm grasp of the concepts that were introduced in the preceding discussion.

Exercises

1. Given a non-empty subset B of a poset A, the element x (not necessarily contained in B) is said to be a *least upper bound* of B if x is an upper bound of B and xPy for every upper bound y of B. Prove that B has at most one least upper bound (so that we may refer to it as *the* least upper bound).

2. The definitions of upper bound, least upper bound, and maximal element may obviously be dualized to furnish definitions of lower bound, greatest lower bound, and minimal element. A

poset A is said to be a *lattice* if every two-member subset of A, $\{a, b\}$, has a least upper bound, denoted $a \vee b$, and a greatest lower bound, denoted $a \wedge b$. Let a non-empty set U be given, let A be the collection of all subsets of U (including $\emptyset$ and U), and let aPb mean that $a \subseteq b$, $a \in A$, $b \in A$. Show that A is a lattice and describe explicitly $a \vee b$ and $a \wedge b$. (Note that A is not linearly ordered unless U consists of a single point.)

3. A lattice A is said to be *complete* if every non-empty subset of A has a greatest lower bound and a least upper bound. Show that the lattice defined in the preceding exercise is complete. On the other hand, if U is an infinite set and A consists of the *finite* subsets of U (with P defined in the same manner), show that A is still a lattice, but is no longer complete.

4. In any lattice, show that $a \vee (b \vee c) = (a \vee b) \vee c$ and $a \wedge (b \wedge c) = (a \wedge b) \wedge c$.

5. Prove or disprove: In any lattice, $a \vee (b \wedge c) = (a \vee b) \wedge (a \vee c)$. (This *is* true for the lattice considered in Exercise 2, but it might nevertheless be false in some other lattice.)

6. Prove, by use of Zorn's Lemma, the extension of Theorem 1–4 of Chapter 4 indicated in the accompanying footnote.

Note: The remaining exercises in this set refer to a Banach space B and the set $\mathscr{B}(B)$ of all operators on B. These exercises should be attempted after the reader has studied Chapter 6. The ideas discussed in these problems can be formulated in a more general setting (Banach algebras), but for simplicity we confine attention to the particular case of $\mathscr{B}(B)$.

7. A non-empty linear manifold $\mathscr{I}$ of $\mathscr{B}(B)$ is termed an *ideal* (in $\mathscr{B}(B)$) if for every member S of $\mathscr{B}(B)$ and every member T of $\mathscr{I}$ the operators ST and TS belong to $\mathscr{I}$. (Clearly, $\mathscr{B}(B)$ and $\{O\}$ are ideals; they are called the *trivial ideals* of $\mathscr{B}(B)$. Any ideal other than $\mathscr{B}(B)$ is called a *proper ideal*; note that $\{O\}$ is a proper ideal.) Prove that a proper ideal contains no invertible members of $\mathscr{B}(B)$.

8. Recalling that $\mathscr{B}(B)$ becomes a Banach space when normed in the manner explained in Chapter 6, we can form the closure $\overline{\mathscr{I}}$ of any ideal $\mathscr{I}$. Prove that $\overline{\mathscr{I}}$ is also an ideal and that if $\mathscr{I}$ is proper, so is $\overline{\mathscr{I}}$.

9. An ideal $\mathscr{I}$ is said to be *maximal* if it is proper and if it is not a proper subset of any other proper ideal. Prove that a maximal ideal must be closed.

10. Prove that at least one maximal ideal exists; in fact, show that any proper non-maximal ideal is contained in at least one maximal ideal. (Hint: Introduce a suitable partial ordering into the class of all proper ideals and show that the hypothesis of Zorn's lemma is satisfied.)

B. CONCERNING THE SPECTRUM OF AN OPERATOR ON A COMPLEX BANACH SPACE

Let T be any operator on the *complex* Banach space B. We wish to show that $\sigma(T)$ is not empty—i.e., that there exists at least one complex number μ such that $(\mu I - T)^{-1}$ does not exist. Suppose, on the contrary, that $(\mu I - T)^{-1}$ exists for all μ. Then, in particular, $(-T)^{-1}$ exists, and so, if f is any non-zero vector, chosen once and for all, $(-T)^{-1}f$ is a well-defined non-zero vector, and so there exists a bounded linear functional l on B such that the function $g(\mu) = l((\mu I - T)^{-1}f)$ is defined for all complex numbers μ and does not vanish when $\mu = 0$. Now, for $|\mu| > \|T\|$ we may write

$$(\mu I - T)^{-1} = \frac{1}{\mu} I + \frac{T}{\mu^2} + \frac{T^2}{\mu^3} + \cdots,$$

and so (cf. Exercise 1)

$$\begin{aligned} g(\mu) &= l\left(\frac{1}{\mu} f + \frac{1}{\mu^2} Tf + \frac{1}{\mu^3} T^2 f + \cdots\right) \\ &= \frac{1}{\mu} l(f) + \frac{1}{\mu^2} l(Tf) + \frac{1}{\mu^3} l(T^2 f) + \cdots. \end{aligned}$$

Therefore,

$$\begin{aligned} |g(\mu)| &\leqslant \frac{1}{|\mu|} |l(f)| + \frac{1}{|\mu|^2} |l(Tf)| + \frac{1}{|\mu|^3} |l(T^2 f)| + \cdots \\ &\leqslant \|l\| \left\{\frac{\|f\|}{|\mu|} + \frac{\|T\| \cdot \|f\|}{|\mu|^2} + \frac{\|T^2\| \cdot \|f\|}{|\mu|^3} + \cdots\right\} \\ &\leqslant \frac{\|l\| \cdot \|f\|}{|\mu|} \left\{1 + \frac{\|T\|}{|\mu|} + \frac{\|T\|^2}{|\mu|^2} + \cdots\right\} \\ &= \frac{\|l\| \cdot \|f\|}{|\mu| - \|T\|}. \end{aligned}$$

Thus, $|g(\mu)|$ approaches zero uniformly as $|\mu| \to \infty$.

Next we observe that for any complex numbers δ and μ,

$$\begin{aligned} g(\mu + \delta) - g(\mu) &= l(((\mu + \delta)I - T)^{-1}f) - l((\mu I - T)^{-1}f) \\ &= l(((\mu + \delta)I - T)^{-1}f - (\mu I - T)^{-1}f) \\ &= l(\{(\mu I - T)(I - \delta(\mu I - T)^{-1})\}^{-1}f - (\mu I - T)^{-1}f). \end{aligned}$$

Fixing μ, let $0 < |\delta| < \|(\mu I - T)^{-1}\|$. Then we may, by Theorem 7–3 of Chapter 6, write

$$\begin{aligned} &g(\mu + \delta) - g(\mu) \\ &= l(\{(\mu I - T)^{-1} + \delta(\mu I - T)^{-2} + \delta^2(\mu I - T)^{-3} + \cdots\}f - (\mu I - T)^{-1}f) \\ &= \delta l((\mu I - T)^{-2}f) + \delta^2 l((\mu I - T)^{-3}f) + \cdots. \end{aligned}$$

Hence,

$$\begin{aligned} \left| \frac{g(\mu + \delta) - g(\mu)}{\delta} - l((\mu I - T)^{-2}f) \right| &\leqslant \|l\| \sum_{n=1}^{\infty} |\delta|^n \, \|(\mu I - T)^{-1}\|^{2+n} \, \|f\| \\ &= \frac{\|l\| \cdot \|f\| \cdot \|(\mu I - T)^{-1}\|^3 \cdot |\delta|}{1 - |\delta| \cdot \|(\mu I - T)^{-1}\|}. \end{aligned}$$

Thus, $\lim_{\delta \to 0} ((g(\mu + \delta) - g(\mu))/\delta)$ exists for every μ (and equals $l(\mu I - T)^{-2}f)$). Therefore $g(\mu)$ is an entire function which approaches zero as $|\mu| \to \infty$. By Liouville's theorem, $g(\mu) \equiv 0$, contrary to the assumption that $g(0) \neq 0$. Hence, $\sigma(T)$ cannot be vacuous.

Note that, in assuming the existence of a non-zero vector f, we have tacitly excluded the trivial case that the space B consists only of the zero vector.

Exercise

1. Justify the term-by-term application of l to the various infinite sums which appear in the argument.

C. THE STIELTJES INTEGRAL

Let f and g be real-valued functions defined on the finite closed interval $[a, b]$, f being continuous and g non-decreasing (but not necessarily continuous). Let Π be any partition of $[a, b]$—i.e., an ordered set of numbers $\{x_0, x_1, x_2, \ldots, x_n\}$ such that $a = x_0 < x_1 < x_2 < \cdots < x_n = b$, $\|\Pi\|$, the *norm* of Π, being defined as the largest of the numbers $x_k - x_{k-1}$, $1 \leqslant k \leqslant n$. Let numbers $\xi_1, \xi_2, \ldots, \xi_n$ be chosen arbitrarily in the intervals $[x_0, x_1], [x_1, x_2], \ldots, [x_{n-1}, x_n]$ respectively. The Riemann-Stieltjes sum $\sum_{k=1}^{n} f(\xi_k)(g(x_k) - g(x_{k-1}))$ is then formed. Since f is uniformly continuous, it is possible, given any positive number ϵ, to find a positive number δ such that $|f(x) - f(x')| < \epsilon/(g(b) - g(a))$ whenever $|x - x'| < \delta$. Hence, whenever $\|\Pi\|$ is less than δ, replacing the numbers $\xi_1, \xi_2, \ldots, \xi_n$ by another admissible set of numbers $\xi_1', \xi_2', \ldots, \xi_n'$ alters the Riemann-Stieltjes sum by less than ϵ, for the absolute value of the difference of the two sums is at most $\sum_{k=1}^n |f(\xi_k') - f(\xi_k)| \, (g(x_k) - g(x_{k-1}))$, which is less than $(\epsilon/(g(b) - g(a))) \sum_{k=1}^n (g(x_k) - g(x_{k-1})) = (\epsilon/(g(b) - g(a)))(g(b) - g(a)) = \epsilon$. (The monotonicity of g is employed here, of course.) Then, exactly as in the development of the theory of the Riemann integral, it follows that if we choose a sequence of partitions whose norms approach zero and form a corresponding Riemann-Stieltjes sum for each of the partitions, the sums converge to a limit which is independent of the choice of the partitions and of the ξ's within each subinterval of each partition. This limit is thus completely determined by the functions f and g, and is denoted $\int_a^b f\,dg$. (In particular, if $g(x) \equiv x$, $\int_a^b f\,dg$ obviously coincides with the Riemann integral $\int_a^b f(x)\,dx$.)

Next, let g_1 and g_2 be two non-decreasing functions. It is easily seen that

$$\int_a^b f\,d(g_1 + g_2) = \int_a^b f\,dg_1 + \int_a^b f\,dg_2. \tag{C–1}$$

If h is expressible as the difference of two non-decreasing functions, say $h = g_1 - g_2$, and also as the difference of two other non-decreasing functions, $h = \tilde{g}_1 - \tilde{g}_2$, we obtain from ($C$–1) the equality

$$\int_a^b f\,dg_1 + \int_a^b f\,d\tilde{g}_2 = \int_a^b f\,d\tilde{g}_1 + \int_a^b f\,dg_2. \tag{C–2}$$

From (C–2) we obtain, in turn,

$$\int_a^b f\,dg_1 - \int_a^b f\,dg_2 = \int_a^b f\,d\tilde{g}_1 - \int_a^b f\,d\tilde{g}_2. \tag{C–3}$$

Thus, we are justified in denoting the common value of both sides of (C–3) as $\int_a^b f\,dh$. The reader is perhaps acquainted with the fact that the (real-valued) function h is expressible as the difference of two non-decreasing functions iff h is of *bounded variation*—by this we mean that there exists a positive number C such that for every partition Π of the interval $[a, b]$ the sum $\sum_{k=1}^n |h(x_k) - h(x_{k-1})|$ does not exceed C.

Now let f and h be complex-valued functions, $f = f_1 + if_2$, $h = h_1 + ih_2$, and suppose furthermore that f_1 and f_2 are both continuous and that h_1 and h_2 are both of bounded variation. Then, as is to be expected, we define $\int_a^b f\,dh$ as follows:

$$\int_a^b f\,dh = \int_a^b f_1\,dh_1 - \int_a^b f_2\,dh_2 + i\int_a^b f_1\,dh_2 + i\int_a^b f_2\,dh_1. \tag{C–4}$$

Finally, it is almost trivial to show that Stieltjes integration is linear—i.e., if f and F are continuous and h and H satisfy the preceding conditions, then for any scalars $\alpha_1, \alpha_2, \beta_1, \beta_2$ the integral $\int_a^b (\alpha_1 f + \alpha_2 F)\, d(\beta_1 h + \beta_2 H)$ is well defined and satisfies the equality

$$\int_a^b (\alpha f_1 + \alpha_2 F)\, d(\beta_1 h + \beta_2 H) = \alpha_1\beta_1 \int_a^b f\, dh$$
$$+ \alpha_1\beta_2 \int_a^b f\, dH + \alpha_2\beta_1 \int_a^b F\, dh + \alpha_2\beta_2 \int_a^b F\, dH. \quad (C\text{–}5)$$

In particular, it should be noted that if h is a step-function having discontinuities only at the points $\eta_1, \eta_2, \ldots, \eta_n$ *inside* the interval $[a, b]$, then $\int_a^b f\, dh$ assumes the simple form $\sum_{k=1}^n f(\eta_k) \cdot \{h(\eta_k + 0) - h(\eta_k - 0)\}$.

Exercise

1. Suppose that both f and h are real-valued, continuous, and non-decreasing. Prove that

$$\int_a^b f\, dh + \int_a^b h\, df = f(b)h(b) - f(a)h(a).$$

D. THE WEIERSTRASS APPROXIMATION THEOREM AND APPROXIMATION BY TRIGONOMETRIC POLYNOMIALS

The Weierstrass theorem is one of the most important approximation theorems in classical analysis, and it has served as a point of departure for vast generalizations. However, we shall present only the original version of the theorem, and out of the vast number of distinct proofs which are known we shall present one which admits an almost obvious generalization to the multi-dimensional case. (Cf. Exercise 1).

Theorem 1: *Let f be a (real-valued or complex-valued) continuous function defined on a compact interval $[a, b]$. Given any positive number ϵ, there exists a polynomial p such that the inequality $|p(x) - f(x)| < \epsilon$ holds throughout the indicated interval.*

PROOF: First we consider the particular case that $f(a) = f(b) = 0$. We extend the function f to all of R by defining $f(x)$ to be zero outside the interval $[a, b]$. Clearly, the extended function thus defined is uniformly continuous, despite the fact that R is not compact. For any positive number t we define the function f_t on all of R as follows.

$$f_t(x) = c(t) \int_{-\infty}^{\infty} f(\xi) e^{-(\xi - x)^2/t}\, d\xi = c(t) \int_a^b f(\xi) e^{-(\xi - x)^2/t}\, d\xi, \quad (D\text{–}1)$$

where $c(t) = (\int_{-\infty}^{\infty} e^{-\xi^2/t}\, d\xi)^{-1} = (\pi t)^{-1/2}$. Since, from the definition of $c(t)$, the equality

$$f(x) = c(t)\int_{-\infty}^{\infty} f(x)e^{-(\xi-x)^2/t}\, d\xi$$

holds for all x, we obtain

$$f_t(x) - f(x) = c(t)\int_{-\infty}^{\infty} (f(\xi) - f(x))e^{-(\xi-x)^2/t}\, d\xi. \tag{D–2}$$

Taking account of the uniform continuity of f, we can choose $\delta(\epsilon)$ such that $|f(\xi) - f(x)| < \epsilon/3$ whenever $|\xi - x| < \delta$. From (D–2) we obtain

$$|f_t(x) - f(x)| \leqslant c(t)\int_{x-\delta}^{x+\delta} \frac{\epsilon}{3} e^{-(\xi-x)^2/t}\, d\xi + c(t)\int_{|\xi-x|\geqslant\delta} 2Me^{-(\xi-x)^2/t}\, d\xi. \tag{D–3}$$

where $M = \max_{x\in R} |f(x)| = \max_{a\leqslant x\leqslant b} |f(x)|$. From ($D$–3) we easily obtain the further inequality

$$\begin{aligned} |f_t(x) - f(x)| &< c(t)\int_{-\infty}^{\infty} \frac{\epsilon}{3} e^{-(\xi-x)^2/t}\, d\xi + 4Mc(t)\int_{\xi=x+\delta}^{\infty} e^{-(\xi-x)^2/t}\, d\xi \\ &= \frac{\epsilon}{3} + 4Mc(t)t^{1/2}\int_{\delta/t^{1/2}}^{\infty} e^{-u^2}\, du \qquad (D\text{–}4) \\ &= \frac{\epsilon}{3} + \frac{4M}{\pi^{1/2}}\int_{\delta/t^{1/2}}^{\infty} e^{-u^2} du. \end{aligned}$$

We now choose t so large that the inequality

$$\frac{4M}{\pi^{1/2}}\int_{\delta/t^{1/2}}^{\infty} e^{-u^2}\, du < \frac{\epsilon}{3}$$

holds; for this choice of t we then have, for all x, the inequality

$$|f_t(x) - f(x)| < \frac{2\epsilon}{3}. \tag{D–5}$$

Referring to (D–1) and confining x to the interval $[a, b]$, we obtain

$$f_t(x) = c(t)\int_a^b f(\xi)\left\{\sum_{k=0}^{N}\frac{(-1)^k(\xi - x)^{2k}}{t^k k!} + \sum_{k=N+1}^{\infty}\frac{(-1)^k(\xi - x)^{2k}}{t^k k!}\right\} d\xi. \tag{D–6}$$

Taking account of the fact that when ξ and x are both confined to the interval $[a, b]$ the inequality $|\xi - x| \leqslant b - a$ must hold, we obtain from

$(D$–$6)$ the inequality

$$\left| f_t(x) - c(t)\int_a^b f(\xi)\left\{\sum_{k=0}^{N}\frac{(-1)^k(\xi - x)^{2k}}{t^k k!}\right\} d\xi \right| \leqslant Mc(t)(b - a)\sum_{k=N+1}^{\infty}\frac{(b - a)^{2k}}{t^k k!}. \quad (D\text{–}7)$$

Now we choose N so large that the right side of $(D$–$7)$ (which is independent of x) is less than $\epsilon/3$. Then from $(D$–$5)$ and $(D$–$7)$ we obtain

$$\left| f(x) - c(t)\int_a^b f(\xi)\ \sum_{k=0}^{N}\frac{(-1)^k(\xi - x)^{2k}}{t^k k!}\right\} d\xi \ \right| \leqslant |f(x) - f_t(x)|$$
$$+ \left| f_t(x) - c(t)\int_a^b f(\xi)\left\{\sum_{k=0}^{N}\frac{(-1)^k(\xi - x)^{2k}}{t^k k!}\right\} d\xi \right| < \frac{2\epsilon}{3} + \frac{\epsilon}{3} = \epsilon. \quad (D\text{–}8)$$

Now we observe that

$$c(t)\int_a^b f(\xi)\left\{\sum_{k=0}^{N}\frac{(-1)^k(\xi - x)^{2k}}{t^k k!}\right\} d\xi$$

is a polynomial (of degree not exceeding $2N$), and so the proof is complete for the case that $f(a) = f(b) = 0$.

In the general case, we may obviously find scalars α and β such that $f(x) - \alpha x - \beta$ vanishes at a and b. We then determine a polynomial q such that $|\{f(x) - \alpha x - \beta\} - q(x)| < \epsilon$ for all x in $[a, b]$. Then the polynomial $\alpha x + \beta + q(x)$ differs from $f(x)$ by less than ϵ throughout the prescribed interval.

Now we turn to a similar theorem, in which we deal with approximation by trigonometric polynomials.

Theorem 2: *Let f be a (real-valued or complex-valued) continuous function defined on the closed interval* $[0, 1]$, *and suppose that* $f(0) = f(1)$. *Then for any positive number ϵ there exists a trigonometric polynomial (i.e., a function expressible in the form $t(x) = \sum_{k=-N}^{N} \gamma_k\, e^{-2\pi ikx}$ for some positive integer N and scalars $\gamma_{-N}, \gamma_{-N+1}, \ldots, \gamma_{N-1}, \gamma_N$) such that the inequality $|f(x) - t(x)| < \epsilon$ holds everywhere in* $[0, 1]$.

Proof: For each positive integer n let the non-negative function K_n be defined as follows: $K_n(x) = c_n((1 + \cos 2\pi x)/2)^n$, where the constant c_n is chosen so that $\int_0^1 K_n = 1$. Let

$$t_n(x) = \int_0^1 f(y)K_n(x - y)\, dy.$$

Upon replacing $\cos 2\pi(x - y)$ in this integral by $\frac{1}{2}(e^{2\pi i(x-y)} + e^{-2\pi i(x-y)})$ one sees immediately that t_n is a trigonometric polynomial. Following closely the proof of the preceding theorem, we write

$$f(x) - t_n(x) = \int_0^1 \{f(x) - f(y)\} K_n(x - y)\, dy. \qquad (D\text{–}9)$$

For convenience we extend the definition of f to all of R as a continuous function by the periodicity condition, $f(x + 1) \equiv f(x)$. (It is at this point we use the condition $f(0) = f(1)$.) Similarly, we consider K_n extended periodically to all of R. The integration appearing in (D–9) may then be performed over any interval of length one; in particular, we choose the interval $[x - \frac{1}{2}, x + \frac{1}{2}]$. Since f is *uniformly* continuous we can choose a positive number δ (independent of x and less than $\frac{1}{2}$) such that

$$|f(x) - f(y)| < \epsilon/2$$

whenever $|x - y| \leqslant \delta$. Then from (D–9) we obtain (taking account of the evenness and non-negativity of K_n and using the symbol M to denote $\max |f|$)

$$\begin{aligned} |f(x) - t_n(x)| &< \frac{\epsilon}{2} \int_{x-\delta}^{x+\delta} K_n(x - y)\, dy + 4M \int_{x+\delta}^{x+1/2} K_n(x - y)\, dy \\ &< \frac{\epsilon}{2} \int_{-1/2}^{1/2} K_n(u)\, du + 4M \int_{\delta}^{1/2} K_n(u)\, du \\ &< \frac{\epsilon}{2} + 4M \cdot \frac{1}{2} \cdot \max_{\delta \leq u \leq 1/2} K_n(u). \end{aligned} \qquad (D\text{–}10)$$

Now, $\max_{\delta \leqq u \leqq \frac{1}{2}} K_n(u) = c_n((1 + \cos 2\pi\delta)/2)^n$, and we obtain an upper bound on c_n as follows:

$$\begin{aligned} 1 &= 2c_n \int_0^{1/2} \left(\frac{1 + \cos 2\pi u}{2}\right)^n du \\ &> 2c_n \int_0^{1/2} \left(\frac{1 + \cos 2\pi u}{2}\right)^n \sin 2\pi u\, du = \frac{2c_n}{\pi(n + 1)}. \end{aligned} \qquad (D\text{–}11)$$

Hence, $c_n < \pi(n + 1)/2$, and so from (D–10) we obtain

$$|f(x) - t_n(x)| < \frac{\epsilon}{2} + M\pi(n + 1) \left(\frac{1 + \cos 2\pi\delta}{2}\right)^n. \qquad (D\text{–}12)$$

Since $(n + 1)((1 + \cos 2\pi\delta)/2)^n$ approaches zero as n increases, we can choose a particular value of n such that $|f(x) - t_n(x)| < \epsilon$.

We conclude with the remark that the kernels K_n which we have employed may be replaced in the preceding argument by many others. In particular, the *Fejer kernels*, defined by the equations

$$F_n(x) = \frac{1}{2n+1}\left\{\frac{\sin(2n+1)\pi x}{\sin \pi x}\right\}^2,$$

play an especially important role in the theory of Fourier series. The reader who is not acquainted with Fejer's discovery of the remarkable properties of these kernels will find a presentation of this topic in most books on Fourier series.

Exercises

1. Let f be continuous in the closed rectangle $a_i \leqslant x_i \leqslant b_i$, $i = 1, 2, \ldots, n$, of R^n. Prove that for any positive number ϵ there exists a polynomial p in the variables $x_1, x_2, \ldots, x_n$ such that the inequality $|f - p| < \epsilon$ holds everywhere in the aforementioned region.

2. Suppose that f is defined in the interval $[a, b]$ and possesses continuous derivatives up to and including the k-th order throughout this interval. Prove that for any positive number ϵ there exists a polynomial p such that the $k + 1$ inequalities $|f^{(j)}(x) - p^{(j)}(x)| < \epsilon, j = 0, 1, 2, \ldots, k$, hold throughout the given interval.

E. THE STRUCTURE OF OPEN SETS OF REAL NUMBERS

In this appendix we shall prove the assertion made in §2–1 concerning the structure of open subsets of R. Let O be any open subset of R. If we set aside the trivial case that $O = \emptyset$, the set O contains a countable infinity $r_1, r_2, r_3, \ldots$ of rational numbers. For each index k we can find (by the *definition* of an open set in R) real numbers a and b such that $r_k \in (a, b) \subset O$. Let a_k equal the greatest lower bound of all values of a and let b_k equal the least upper bound of all values of b for which the preceding conditions (namely $r_k \in (a, b) \subset O$) are satisfied. (Note that we must allow the possibilities $a_k = -\infty$ and $b_k = +\infty$. Also, in the case that $a_k[b_k]$ is finite we are exploiting the fact that every non-empty set of real numbers which is bounded below[above] possesses a greatest lower[least upper] bound.) Let I_k be the open interval (a_k, b_k). Evidently $I_k \subseteq O$, and so $\bigcup_{k=1}^{\infty} I_k \subseteq O$. In order to show that $\bigcup_{k=1}^{\infty} I_k = O$, we first observe that $r_k \in I_k$, so that $\bigcup_{k=1}^{\infty} I_k$ contains all the rational numbers contained in O. Let s be any *irrational* number contained in O. Then, since O is open, we can find an open interval (a, b) such that $a < s < b$ and $(a, b) \subset O$. Since the rational numbers form a dense subset of R, we can find an index j such that $r_j \in (a, b)$. It follows that $s \in I_j \subseteq \bigcup_{k=1}^{\infty} I_k$. Thus, $\bigcup_{k=1}^{\infty} I_k = O$.

Now, it is evident, from the manner in which the intervals $I_1, I_2, I_3, \ldots$ are defined, that either $I_j = I_k$ or $I_j \cap I_k = \emptyset$. Thus, O has been expressed as the union of a finite or countably infinite collection of disjoint open intervals.

Finally, suppose that O is expressed as a union of disjoint, open, non-empty intervals, say $O = \bigcup_\alpha J_\alpha$. (Note that we are not assuming that the collection $\{J_\alpha\}$ is finite or countably infinite.) For any index α, J_α must contain a rational member of O, say r_m (since J_α is a *non-empty* open interval). It then follows, from the manner in which the I_k's were constructed, that $J_\alpha \subseteq I_m$. If J_α were a *proper* subset of I_m, at least one of the two end-points of J_α, say y, would lie in I_m. Since J_α is open, $y \notin J_\alpha$, and so y must belong to $J_{\tilde{\alpha}}$ for some index $\tilde{\alpha}$ different from α; but then, since $J_{\tilde{\alpha}}$ is open, the intervals J_α and $J_{\tilde{\alpha}}$ would overlap, contrary to hypothesis. Thus, each J_α must coincide with one of the intervals I_m, and so we have shown that the decomposition of O into disjoint non-empty open intervals is unique (aside from order).

F. INFINITE SERIES AND THE NUMBER SYSTEM $[0, +\infty]$

Given any sequence $a_1, a_2, a_3, \ldots$ of real numbers, by the *series* $a_1 + a_2 + a_3 + \cdots$ we mean the sequence of partial sums $s_1, s_2, s_3, \ldots$, where $s_n = \sum_{k=1}^{n} a_k$. The series is said to be *convergent* if the sequence of partial sums is convergent; when this occurs, the sum of the series is defined to be the number $\lim_{n\to\infty} s_n$. If the series is not convergent, it is said to be *divergent*.

If the terms of the series $a_1 + a_2 + a_3 + \cdots$ are all non-negative, then the corresponding sequence of partial sums is monotone non-decreasing, and from the fundamental principles of real analysis it follows that the series converges iff the set of partial sums is bounded above; for example, the series $1/1^2 + 1/2^2 + 1/3^2 + \cdots$ is convergent, since it is easily demonstrated that all the partial sums satisfy the inequality $s_n < 2$. (Proof: For $n > 1$,

$$s_n < \frac{1}{1^2} + \frac{1}{1\cdot 2} + \frac{1}{2\cdot 3} + \cdots + \frac{1}{(n-1)n}$$

$$= 1 + \left(\frac{1}{1} - \frac{1}{2}\right) + \left(\frac{1}{2} - \frac{1}{3}\right) + \cdots + \left(\frac{1}{n-1} - \frac{1}{n}\right)$$

$$= (1+1) + \left(-\frac{1}{2} + \frac{1}{2}\right) + \left(-\frac{1}{3} + \frac{1}{3}\right) + \cdots + \left(-\frac{1}{n-1} + \frac{1}{n-1}\right) - \frac{1}{n}$$

$$= 2 - \frac{1}{n} < 2.\Big)$$

It is a remarkable and important fact that if a series consisting of non-negative terms is convergent, any rearrangement of this series (we shall not give a formal definition of a rearrangement—the meaning of this expression should be clear) is also convergent, and furthermore the original series and the new series have the same sum. Actually, this result, remarkable enough in itself, can be extended considerably; we shall content ourselves here with stating the fact that if the original series is decomposed into a number of series (even infinitely many), each of the series of the latter collection is convergent, and the sum of their respective sums agrees with the sum of the original series. For example, let us break up the convergent series $a_1 + a_2 + a_3 + \cdots$ into infinitely many series, $S_1, S_2, S_3, \ldots$, in the following way.

$$S_1\colon\ a_1 + a_2 + a_4 + a_7 + a_{11} + \cdots,$$

$$S_2\colon\ a_3 + a_5 + a_8 + a_{12} + \cdots,$$

$$S_3\colon\ a_6 + a_9 + a_{13} + \cdots,$$

$$S_4\colon\ a_{10} + a_{14} + \cdots,$$

$$S_5\colon\ a_{15} + \cdots,$$

$$S_6\colon\ a_{21} + \cdots,$$

and so forth. Then each of the series $S_1, S_2, S_3, \ldots$ is convergent, and if we denote their sums by $\sigma_1, \sigma_2, \sigma_3, \ldots$ respectively, then the series $\sigma_1 + \sigma_2 + \sigma_3 + \cdots$ is convergent, and its sum equals the sum of the original series, $a_1 + a_2 + a_3 + \cdots$.

Next, we define a series of real, but not necessarily non-negative, numbers $b_1 + b_2 + b_3 + \cdots$ to be *absolutely convergent* if the series $|b_1| + |b_2| + |b_3| + \cdots$ is convergent. Note that we do not include in the definition the hypothesis that the original series is convergent—it is a *theorem* that an *absolutely convergent* series, as defined, is convergent, and that the assertion concerning the rearrangement of a convergent series of non-negative numbers holds true in the present case also, as does the assertion concerning the breakup of a series into a collection of series.

The situation changes radically if the series is convergent but not absolutely convergent; when this happens, the series is said to be *conditionally convergent*. For example, the series $1 - \frac{1}{2} + \frac{1}{3} - \frac{1}{4} + \cdots$ is not absolutely convergent, since the series $1 + \frac{1}{2} + \frac{1}{3} + \frac{1}{4} + \cdots$ is known to be divergent, but it *is* convergent; this is guaranteed by the alternating series test, with which the reader is presumably familiar. Furthermore, by the test just cited, it is known that the sum of this series is less than s_1 $(= 1)$ and more than s_2 $(= \frac{1}{2})$; in fact, the sum is $\log 2$ $(= 0.693\cdots)$. However, we shall need only the fact that the sum of this series, which we shall denote by σ, is not zero. The series $\frac{1}{2} - \frac{1}{4} + \frac{1}{6} - \frac{1}{8} + \cdots$ then converges to $\frac{1}{2}\sigma$, as does the series $0 + \frac{1}{2} + 0 - \frac{1}{4} + 0 + \frac{1}{6} + 0 - \frac{1}{8} + \cdots$. (We do not prove here the elementary fact that when a convergent series is multiplied by a fixed factor, the new series is also convergent and its sum is equal to the product of the fixed factor and the sum of the original series, nor the fact that the convergence of a series and the value of its sum are unaffected by inserting any finite number of zeros between any pairs of consecutive terms or by suppressing any zeros which may appear; also, we shall accept the fact that two convergent series, when added termwise, furnish a convergent series whose sum is obtained by adding the sums of the two given series.) Thus, the series $1 + 0 + \frac{1}{3} - \frac{1}{2} + \frac{1}{5} + 0 + \frac{1}{7} - \frac{1}{4} + \frac{1}{9} + 0 + \frac{1}{11} - \frac{1}{6} + \cdots$, and hence the series $1 + \frac{1}{3} - \frac{1}{2} + \frac{1}{5} + \frac{1}{7} - \frac{1}{4} + \frac{1}{9} + \frac{1}{11} - \frac{1}{6} + \cdots$, must converge to $3\sigma/2$. Now it is readily seen that this series is a rearrangement of the original series, obtained by taking the first two positive terms, the first negative term, the next two positive terms, the second negative term, and so forth. Since $\sigma \neq 0$, we see that the sum of the original series has been altered by this rearrangement. This example (which illustrates a very remarkable theorem which is developed in the accompanying exercises) shows that conditional convergence is a much more fragile phenomenon than absolute convergence.

Returning now to series of non-negative terms, we remark that it is often convenient to assign a sum to such a series even when it is not convergent. (In particular, as is evident in Chapter 2, this is very desirable in developing the theory of Lebesgue measure.) We therefore extend R^+, the non-negative half of the real number system, by adding to it one new member, denoted $+\infty$ (the $+$ sign is often omitted), having the following properties: (*i*) $(+\infty) + a = +\infty$, where a is either $+\infty$ or a member of R^+; (*ii*) $a \cdot (+\infty) = +\infty$ if $a = +\infty$ or if a is any non-zero member of R^+, but $0 \cdot (+\infty) = 0$; (*iii*) $(+\infty) - a = +\infty$ if a is any member of R^+; (*iv*) $a < +\infty$ (or $+\infty > a$) for any member a of R^+; (*v*) $(+\infty) - (+\infty)$ is *undefined.* We denote this enlargement of R^+ by the symbol $[0, +\infty]$ (in contrast to $[0, +\infty)$, which is the same as R^+).

We then *assign* to any divergent series whose terms belong to R^+ as a sum the number $+\infty$, and we also assign this value as the sum of any series whose terms are members of $[0, +\infty]$ provided that at least one term of the series is $+\infty$. It then becomes almost self-evident that the theorems concerning convergent series of non-negative real numbers remain valid in the more general class of series which we have just introduced.

Exercises

1. (*a*) Prove that a conditionally convergent series (of real numbers) must contain infinitely many positive terms and infinitely many negative terms.

(*b*) Show that the series consisting of the positive terms of a conditionally convergent series must be divergent (or, in the terminology introduced in the last paragraph, the sum of the series must be $+\infty$). (Of course, a similar result must hold for the series formed from the negative terms of the original series.)

2. Prove the following remarkable theorem, due to Riemann: If the series $a_1 + a_2 + a_3 + \cdots$ is conditionally convergent and if x is any real number, it is possible to rearrange the given series so as to converge to the sum x. Hint: Taking account of (*b*) of the preceding exercise, select enough positive terms from the beginning of the given series so that their sum exceeds x. (We assume here for convenience that $x \geqslant 0$; if $x < 0$, only a trivial modification is needed.) Then subtract from this sum enough negative terms so that a number less than x is obtained. Then return to the remaining positive terms until a sum exceeding x is once again obtained, and so forth. (For simplicity, assume at first that the given series contains no zero terms; these may easily be accounted for in the rearrangement at the very end of the task.)

G. LIMIT SUPERIOR AND LIMIT INFERIOR

Let $a_1, a_2, a_3, \ldots$ be a given sequence of real numbers. For convenience we shall assume at present that this sequence is bounded above and below, but later we shall indicate very briefly how the concepts to be developed here are extended when the given sequence is unbounded in at least one direction.

According to the Bolzano-Weierstrass theorem, the given sequence contains a convergent subsequence; furthermore, if the given sequence is convergent to a number l, then every subsequence also converges to l. On the other hand, if the original sequence does not converge, there exist at least two *distinct* numbers, l_1 and l_2, such that it is possible to extract from the original sequence a subsequence which converges to l_1 and a subsequence which converges to l_2. For example, from the (bounded) sequence $\frac{1}{2}, \frac{3}{2}, \frac{1}{3}, \frac{4}{3}, \frac{1}{4}, \frac{5}{4}, \ldots$ we can extract the subsequences $\frac{1}{2}, \frac{1}{3}, \frac{1}{4}, \ldots$ and $\frac{3}{2}, \frac{4}{3}, \frac{5}{4}, \ldots$ which converge to 0 and 1, respectively; it is not difficult to see that no other numbers can be the limits of subsequences of the given sequence. Incidentally, the simpler sequence $1, 0, 1, 0, 1, 0, \ldots$ serves to illustrate an important fact, namely, that the set of all possible limits of convergent subsequences of a given sequence does not in general coincide with the set of limit-points of the set of numbers appearing in the sequence, for the set consisting exclusively of the numbers 0 and 1, being finite, has no limit-points. (On the other hand, as pointed out in Chapter 1, any limit-point of the set of numbers appearing in any sequence is the limit of some suitably chosen subsequence.)

Thus, given any bounded sequence $a_1, a_2, a_3, \ldots$ of real numbers, the set L of all possible limits of convergent subsequences is non-empty; furthermore (cf. Exercise 1) L is closed. Since $L \subseteq [a, b]$, where $[a, b]$ is any closed interval containing all members of the given sequence, it follows that L must be compact, and so it contains (not merely *has*) a least upper bound and a greatest lower bound. These numbers are known, respectively, as the *upper limit*, or *limit superior*, and the *lower limit*, or *limit inferior*, of the given sequence; the notations $\limsup_{n\to\infty} a_n$, or $\overline{\lim}_{n\to\infty} a_n$, and $\liminf_{n\to\infty} a_n$, or $\underline{\lim}_{n\to\infty} a_n$, are employed. (The notation $n \to \infty$ is really unnecessary and is often omitted.)

We now state the following theorem, which gives two alternative characterizations of the upper and lower limits; the proof is left to the reader as Exercise 6.

Theorem 1: (*a*) *Let $a_1, a_2, a_3, \ldots$ be any bounded sequence of real numbers and let $\overline{\lim}\, a_n$ be denoted by α. Then α is the unique number possessing the following property: For every positive number ϵ, the inequality $a_n < \alpha + \epsilon$ is satisfied for all indices n exceeding some index N (which may depend on ϵ), while the inequality $a_n > \alpha - \epsilon$ is satisfied for infinitely many indices n. (A dual characterization holds for $\underline{\lim}\, a_n$, of course.)*

(*b*) *As in* (*a*), *let $a_1, a_2, a_3, \ldots$ be a bounded sequence of real numbers, and let $s_n = \sup \{a_n, a_{n+1}, a_{n+2}, \ldots\}$. Then the sequence $s_1, s_2, s_3, \ldots$ is monotone non-increasing, bounded below by* $\inf \{a_1, a_2, a_3, \ldots\}$, *and hence convergent to a limit. This limit coincides with* $\overline{\lim}\, a_n$. *(As in* (*a*), *a dual characterization holds for* $\underline{\lim}\, a_n$.)

If the sequence $a_1, a_2, a_3, \ldots$ is unbounded in either one or both directions, it may be impossible to apply the preceding definitions if we restrict ourselves to operating within the real number system R. Without going into details, we mention that the brief explanation provided in §2–4 of the extended real number system R^* makes it evident how to proceed in such a case. We illustrate with a few simple examples:

(*a*) For the sequence $0, 1, 0, 2, 0, 3, \ldots$ the lower limit is 0, the upper limit is $+\infty$.

(*b*) For the sequence $1, -2, 3, -4, 5, -6, \ldots$ the lower limit is $-\infty$, the upper limit is $+\infty$.

(*c*) For the sequence $1, 2, 3, 4, 5, 6, \ldots$ the lower limit and the upper limit are both equal to $+\infty$.

The reader may find it helpful, in mastering the ideas presented, to study the following important theorem.

Theorem 2 (Cauchy-Hadamard): *Let $a_1, a_2, a_3, \ldots$ be any sequence of complex numbers. Then the series $a_1z + a_2z^2 + a_3z^3 + \cdots$ converges for any complex number z satisfying the inequality $|z| < (\overline{\lim}\, |a_n|^{1/n})^{-1}$ and diverges if $|z| > (\overline{\lim}\, |a_n|^{1/n})^{-1}$. (Of course, if*

$$\overline{\lim}\, |a_n|^{1/n} = 0$$

this is to be interpreted as meaning that the series converges for all values of z, while if $\overline{\lim}\, |a_n|^{1/n} = +\infty$ the conclusion is that the series converges only when $z = 0$.)

PROOF: For convenience we assume that $\overline{\lim}\, |a_n|^{1/n}$ is positive and finite, so that we may express it in the form $1/R$, where $0 < R < +\infty$. (R is known as the *radius of convergence* of the given series; the reason for this terminology is obvious from the theorem.) The validity of the conclusion in the extreme cases $R = 0$, $R = +\infty$ will be apparent from the proof to be presented.

If $|z| < R$, then for any positive number ϵ, the inequality $|a_nz^n|^{1/n} < (1/R + \epsilon)\,|z|$ holds for all sufficiently large n. Since $|z| < R$, ϵ can be chosen so small that $(1/R + \epsilon)\,|z| < 1$. Denoting the left side of this inequality by α, we conclude that for all sufficiently large n the inequality $|a_nz^n| < \alpha^n$ holds. Since the series $\alpha + \alpha^2 + \alpha^3 + \cdots$ converges, the series must converge. (Indeed, the series converges absolutely, and if z is restricted by the further condition $|z| \leqslant R'$, where R' is any number smaller than R, the convergence is uniform.)

Conversely, if $|z| > R$, it is readily seen by examining the preceding paragraph that the inequality $|a_nz^n| > 1$ must hold for infinitely many values of n (but not necessarily for all sufficiently large values of n). Thus, the terms of the series do not approach zero with increasing n; hence the series cannot possibly converge.

As can be shown by simple examples (cf. Exercise 7), the series may converge for some values, no values, or all values of z satisfying the condition $|z| = R$.

Exercises

1. Prove that the set L defined previously is closed.

2. Demonstrate a sequence whose terms all lie in the *open* interval $(0, 1)$ and whose set L consists of all points of the *closed* interval $[0, 1]$.

3. Let the sequence $b_1, b_2, b_3, \ldots$ be a rearrangement of the sequence $a_1, a_2, a_3, \ldots$. Prove that the sets L associated with these two sequences are identical, so that, in particular, $\overline{\lim}\, a_n = \overline{\lim}\, b_n$ and $\underline{\lim}\, a_n = \underline{\lim}\, b_n$.

4. Show that $\underline{\lim}\,(-a_n) = -\overline{\lim}\,a_n$ and similarly that $\overline{\lim}\,(-a_n) = -\underline{\lim}\,a_n$.

5. Show that $\overline{\lim}\,(a_n + b_n) \leqslant \overline{\lim}\,a_n + \overline{\lim}\,b_n$ and similarly that $\underline{\lim}\,(a_n + b_n) \geqslant \underline{\lim}\,a_n + \underline{\lim}\,b_n$.

6. Prove Theorem 1.

7. Consider the series $z + z^2 + z^3 + \cdots$, $z + z^2/2 + z^3/3 + \cdots$, and $z + z^2/2^2 + z^3/3^2 + \cdots$. Show that for each of these series the radius of convergence equals one, and that the first series converges for no value of z whose modulus is one, while the third series converges for all such values and the second converges for at least one such value and diverges for at least one such value.

H. THE FOURIER TRANSFORM IN $L^2(R)$

In this appendix we indicate briefly how the Fourier transform may be interpreted as a unitary operator on $L^2(R)$.

Let f_1 and f_2 be the characteristic functions of the (finite) intervals (a_1, b_1) and (a_2, b_2) respectively. It is not assumed that these intervals are disjoint. The Fourier transforms $\hat{f}_1, \hat{f}_2$ of these functions are defined, for all *real* values of t, as follows:

$$\hat{f}_k(t) = \int_{-\infty}^{\infty} f_k(x)e^{-2\pi itx}\,dx = \int_{a_k}^{b_k} e^{-2\pi itx}\,dx = \frac{e^{-2\pi ia_kt} - e^{-2\pi ib_kt}}{2\pi it}. \qquad (H\text{–}1)$$

By an elementary computation one obtains

$$\hat{f}_1(t)\overline{\hat{f}_2(t)} = \frac{g(t) + ih(t)}{4\pi^2t^2}, \qquad (H\text{–}2)$$

where

$$g(t) = \cos 2\pi(b_2 - b_1)t + \cos 2\pi(a_2 - a_1)t - \cos 2\pi(b_2 - a_1)t - \cos 2\pi(a_2 - b_1)t$$

and

$$h(t) = \sin 2\pi(b_2 - b_1)t + \sin 2\pi(a_2 - a_1)t - \sin 2\pi(b_2 - a_1)t - \sin 2\pi(a_2 - b_1)t.$$

Since $h(t)/t^2$ is an odd function, one concludes from $(H\text{–}2)$ that

$$\int_{-\infty}^{\infty} \hat{f}_1(t)\overline{\hat{f}_2(t)}\,dt = \frac{1}{4\pi^2}\int_{-\infty}^{\infty} \frac{g(t)}{t^2}\,dt,$$

and four uses of the identity $\cos 2v = 1 - 2\sin^2 v$ lead to the equality

$$\int_{-\infty}^{\infty} \hat{f}_1(t)\overline{\hat{f}_2(t)}\,dt = \frac{1}{2\pi^2}\int_{-\infty}^{\infty} \frac{1}{t^2}(\sin^2 \pi(b_2 - a_1)t + \sin^2 \pi(a_2 - b_1)t$$
$$- \sin^2 \pi(b_2 - b_1)t - \sin^2 \pi(a_2 - a_1)t)\,dt. \quad (H\text{–}3)$$

For any real number c, $\int_{-\infty}^{\infty} (\sin^2 ct/t^2)\,dt = \pi\,|c|$ (cf. Exercises 3 and 10), and so (H–3) furnishes the equality

$$\int_{-\infty}^{\infty} \hat{f}_1(t)\overline{\hat{f}_2(t)}\,dt = \tfrac{1}{2}\{|b_2 - a_1| + |a_2 - b_1| - |b_2 - b_1| - |a_2 - a_1|\}. \quad (H\text{–}4)$$

By considering all possible relationships among the quantities a_1, a_2, b_1, b_2 (subject, of course, to the restrictions $a_1 < b_1$ and $a_2 < b_2$) one readily confirms that the right side of (H–4) equals the length of the intersection of the two given intervals. Therefore,

$$\int_{-\infty}^{\infty} \hat{f}_1(t)\overline{\hat{f}_2(t)}\,dt = \int_{-\infty}^{\infty} f_1(x)\overline{f_2(x)}\,dx. \quad (H\text{–}5)$$

We now proceed to extend the class of functions for which (H–5) holds true. If g_1 and g_2 are any step-functions vanishing outside some sufficiently large interval, they may be expressed in the form

$$g_1(x) = \sum_{j=1}^{N_1} \alpha_j f_j^{(1)}(x), \qquad g_2(x) = \sum_{k=1}^{N_2} \beta_k f_k^{(2)}(x), \quad (H\text{–}6)$$

where the functions $f_1^{(1)}, f_2^{(1)}, \ldots, f_{N_1}^{(1)}$, and similarly the functions $f_1^{(2)}$, $f_2^{(2)}, \ldots, f_{N_2}^{(2)}$, are characteristic functions of *disjoint* intervals. Defining the Fourier transforms $\hat{g}_1$ and $\hat{g}_2$ of g_1 and g_2 as

$$\int_{-\infty}^{\infty} g_1(x)e^{-2\pi itx}\,dx \quad \text{and} \quad \int_{-\infty}^{\infty} g_2(x)e^{-2\pi itx}\,dx,$$

respectively, we obtain (taking account of (H–5))

$$\int_{-\infty}^{\infty} \hat{g}_1(t)\overline{\hat{g}_2(t)}\,dt = \sum_{j=1}^{N_1}\sum_{k=1}^{N_2} \alpha_j\bar{\beta}_k \int_{-\infty}^{\infty} \hat{f}_j^{(1)}(t)\overline{\hat{f}_k^{(2)}(t)}\,dt$$
$$= \sum_{j=1}^{N_1}\sum_{k=1}^{N_2} \alpha_j\bar{\beta}_k \int_{-\infty}^{\infty} f_j^{(1)}(x)\overline{f_k^{(2)}(x)}\,dx = \int_{-\infty}^{\infty} g_1(x)\overline{g_2(x)}\,dx. \quad (H\text{–}7)$$

Now let h_1 and h_2 be any two functions belonging to $L^2(R)$ and vanishing outside some sufficiently large interval $[a, b]$. By the Schwarz inequality we obtain $\{\int_{-\infty}^{\infty} |h_k(x)|\, dx\}^2 = \{\int_a^b |h_k(x)|\, dx\}^2 \leqslant \{\int_a^b 1^2\, dx\} \times \{\int_a^b |h_k(x)|^2\, dx\} = (b - a)\, \|h_k\|_2^2 < \infty$, and so we have shown that the functions h_1 and h_2 belong to $L^1(R)$. Since $|e^{-2\pi i t x}| = 1$ for all real values of t and x, the integrals $\int_{-\infty}^{\infty} h_k(x)\, e^{-2\pi i t x}\, dx$ exist for all real t; in fact, the value of the integral depends continuously on t and approaches zero as $|t| \to \infty$. (Cf. Exercise 7.) These functions are, of course, called the *Fourier transforms*, $\hat{h}_1$ and $\hat{h}_2$, of h_1 and h_2. By a suitable approximation argument, which we do not present here, it can be shown that (H–7) generalizes to the present case:

$$\int_{-\infty}^{\infty} \hat{h}_1(t)\overline{\hat{h}_2(t)}\, dt = \int_{-\infty}^{\infty} h_1(x)\overline{h_2(x)}\, dx. \qquad (H\text{–}8)$$

(The integral on the right is meaningful; this is guaranteed by the Schwarz inequality, even without the restriction that the functions h_1 and h_2 vanish outside $[a, b]$.) However, if h_1 and h_2 fail to vanish outside some finite interval the formal definition $\hat{h}_k(t) = \int_{-\infty}^{\infty} h_k(x)\, e^{-2\pi i t x}\, dx$ may be meaningless. In this case we proceed as follows. Let h be any member of $L^2(R)$ and for each positive integer n let $h^{(n)}$ coincide with h in $[-n, n]$ and vanish outside this interval. Then we may construct the sequence of Fourier transforms $\hat{h}^{(1)}, \hat{h}^{(2)}, \hat{h}^{(3)}, \ldots$. Replacing both h_1 and h_2 in (H–8) by $h^{(n)} - h^{(m)}$, we obtain

$$\|\hat{h}^{(n)} - \hat{h}^{(m)}\|_2 = \|h^{(n)} - h^{(m)}\|_2. \qquad (H\text{–}9)$$

Since the sequence $h^{(1)}, h^{(2)}, h^{(3)}, \ldots$ is Cauchy, (H–9) guarantees that the corresponding sequence $\hat{h}^{(1)}, \hat{h}^{(2)}, \hat{h}^{(3)}, \ldots$ is also Cauchy. By the Riesz-Fischer theorem it then follows that there exists a function $\hat{h}$ in $L^2(R)$ such that $\lim_{n\to\infty} \|\hat{h} - \hat{h}^{(n)}\|_2 = 0$, and we now *define* $\hat{h}$ to be the Fourier transform of h. (The function $\hat{h}$ is, of course, uniquely determined up to a null-set; if the integral $\int_{-\infty}^{\infty} h(x)e^{-2\pi i t x}\, dx$ does exist for all t, the corollary to the Riesz-Fischer theorem guarantees the equivalence of the two definitions of $\hat{h}$.) Replacing h_1 and h_2 in (H–8) by the corresponding functions $h_1^{(n)}$ and $h_2^{(n)}$ and letting n increase without bound, we find that (H–8) continues to hold for any pair of functions in $L^2(R)$ and their Fourier transforms.

Thus, the Fourier transform is a linear transformation mapping $L^2(R)$ into itself and preserving inner products. (Observe that (H–8) can be written in the form $(\hat{h}_1, \hat{h}_2) = (h_1, h_2)$. Setting $h_1 = h_2 = h$, we obtain $\|\hat{h}\|_2 = \|h\|_2$.) The question now arises as to whether this transformation maps $L^2(R)$ *onto* itself. The proof that this is indeed so will be sketched briefly; thus, the Fourier transform will be shown to be a unitary operator on $L^2(R)$.

Let f be the characteristic function of the interval (a, b). Then $\hat{f}(t) = (e^{-2\pi i a t} - e^{-2\pi i b t})/2\pi i t$; temporarily replacing (for convenience) the symbol $\hat{f}$ by g, we may express $\hat{g}(t)$ as follows:

$$\hat{g}(t) = \lim_{n\to\infty} \int_{-n}^{n} \frac{e^{-2\pi iax} - e^{-2\pi ibx}}{2\pi ix} e^{-2\pi itx}\,dx. \qquad (H\text{–}10)$$

(Strictly speaking, the limit is to be taken in the sense of the norm on $L^2(R)$, but it will be seen that the limit exists in the ordinary sense; referring again to the corollary to the Riesz-Fischer theorem, one sees that the two limits agree almost everywhere.) A simple computation shows that (*H*–10) may be rewritten in the form

$$\hat{g}(t) = \lim_{n\to\infty} \int_{-n}^{n} \frac{\sin \pi(b-a)x}{x} e^{-\pi i(2t+a+b)x}\,dx. \qquad (H\text{–}11)$$

Since the imaginary part of the integrand is odd, it may be discarded, and so (*H*–11) may be rewritten as follows:

$$\begin{aligned}\hat{g}(t) &= \lim_{n\to\infty} \int_{-n}^{n} \frac{\sin \pi(b-a)x \cos \pi(2t+a+b)x}{\pi x}\,dx \\ &= \frac{1}{2\pi} \lim_{n\to\infty} \left\{ \int_{-n}^{n} \frac{\sin 2\pi(t+b)}{x}\,dx - \int_{-n}^{n} \frac{\sin 2\pi(t+a)x}{x}\,dx \right\}. \qquad (H\text{–}12)\end{aligned}$$

Taking account of Exercises 3 and 10 we obtain

$$\hat{g}(t) = \tfrac{1}{2}\{\operatorname{sgn}(t+b) - \operatorname{sgn}(t+a)\} = \begin{cases} 1 & \text{if } -b < t < -a, \\ \tfrac{1}{2} & \text{if } t = -a \text{ or } t = -b, \\ 0 & \text{otherwise.} \end{cases} \qquad (H\text{–}13)$$

Thus, the equality $\hat{g}(t) = f(-t)$ holds everywhere except at $t = -a$ and at $t = -b$; if we had assigned to $f(a)$ and $f(b)$ the value $\frac{1}{2}$ instead of zero, the equality $g(t) = f(-t)$ would have held without exception. In any case, the result just obtained may be written in the following striking form:

$$\hat{\hat{f}}(t) = f(-t) \quad \text{almost everywhere.} \qquad (H\text{–}14)$$

Replacing t by $-t$, we obtain

$$\hat{\hat{f}}(-t) = f(t) \quad \text{almost everywhere.} \qquad (H\text{–}15)$$

It is evident that (*H*–15) will continue to hold if f is any step-function vanishing outside some finite interval. By an approximation argument similar to that indicated earlier in this appendix, it can be shown that every function in $L^2(R)$ satisfies (H–15). That is to say, given any function f belonging to $L^2(R)$, form its Fourier transform $\hat{f}$ and then form the function h defined by the equation $h(t) = \hat{f}(-t)$. Then $f = \hat{h}$. Thus, every member of $L^2(R)$ is the Fourier transform of some member of $L^2(R)$, and so the Fourier transform is a unitary operator. For emphasis we repeat that, in general, the Fourier transform $\hat{f}$ must be interpreted as the limit in the norm of $L^2(R)$ of the integrals $\int_{-n}^{n} f(x)e^{-2\pi itx}\,dx$, rather than as the integral $\int_{-\infty}^{\infty} f(x)e^{-2\pi itx}\,dx$. However, with the understanding that the latter integral is to be interpreted in the sense explained previously, we observe that (H–15) can be written in the following form, known as the *Fourier inversion formula:*

$$\text{If} \quad g(t) = \int_{-\infty}^{\infty} f(x)e^{-2\pi itx}\,dx, \quad \text{then} \quad f(x) = \int_{-\infty}^{\infty} g(t)e^{2\pi itx}\,dt. \qquad (H\text{–}16)$$

Since, as shown previously (cf. (H–15)), two applications of the Fourier transform convert $f(x)$ into $f(-x)$, it follows that four applications carry the function f into itself. Thus, the Fourier transform is a fourth root of the identity operator.

We conclude with the remark that the Fourier transform is a very powerful tool in many branches of analysis. Very often a difficult problem involving an unknown function f may be converted into a much simpler problem involving $\hat{f}$; by solving the simpler problem and then employing the Fourier inversion formula, one solves the original problem.

Exercises

1. Show that $\sin^2 x/x^2 \in L^1(R)$ and that $\sin x/x \notin L^1(R)$.

2. Show that $\lim_{A\to\infty} \int_0^A (\sin x/x)\,dx$ exists and equals $\int_0^\infty (\sin^2 x/x^2)\,dx$.

3. Let $\int_0^\infty (\sin x/x)\,dx$ (understood to mean the limit appearing in the preceding exercise) be denoted by C. (It will be shown in a later exercise that $C = \pi/2$.) Show that, for any real constant α, $\int_0^\infty (\sin \alpha x/x)\,dx = C \operatorname{sgn} \alpha$, where $\operatorname{sgn} \alpha$, the *signum* of α, is $+1$, -1, or 0 according as $\alpha > 0$, $\alpha < 0$, $\alpha = 0$, respectively. Similarly, show that $\int_0^\infty (\sin^2 \alpha x/x^2)\,dx = C\,|\alpha|$.

4. Let f be any step-function vanishing outside a finite interval. Show that $\hat{f}(t)$ varies continuously with t and approaches zero as $|t| \to \infty$. (It is understood that t assumes only real values.)

5. In the preceding exercise replace "step-function" by "continuous function." Show that the conclusion still holds.

6. Now replace "continuous function" by "summable function," and prove that the conclusion still holds. (Retain the restriction that f vanishes outside a finite interval.)

7. Now show that the conclusion holds for any function in $L^1(R)$, even if it does not vanish outside some finite interval. (This result is known as the Riemann-Lebesgue Lemma; the three preceding exercises should be looked upon as stepping-stones to the proof of this result, which plays a major role in Fourier analysis.)

8. Prove that the function $f(x) = 1/x - 1/(2 \sin \frac{1}{2}x)$ is continuous in $[0, \pi]$ if $f(0) = 0$.

9. Prove that, for any positive integer n,

$$\int_0^\pi \frac{\sin (n + \frac{1}{2})x}{2 \sin \frac{1}{2}x}\, dx = \tfrac{1}{2}\pi.$$

10. Use Exercises 5, 8, and 9 to show that $\int_0^\infty (\sin x/x)\, dx = \pi/2$ (and hence that $\int_{-\infty}^\infty (\sin x/x)\, dx = \pi$).

SOME SUGGESTIONS FOR FURTHER READING

N. I. AKHIEZER AND I. M. GLAZMAN
Theory of Linear Operators in Hilbert Space, Ungar, 1961

S. BANACH
Théorie des Opérations Linéaires, Chelsea, 1955 (Reprint)

S. BERGMAN
The Kernel Function and Conformal Mapping, American Mathematical Society, 1950

R. C. BUCK, *Editor*
Studies in Modern Analysis, Mathematical Association of America, 1962

N. DUNFORD AND J. T. SCHWARTZ
Linear Operators, Interscience, 1958 (Vol. I) and 1963 (Vol. II)

B. EPSTEIN
Orthogonal Families of Analytic Functions, Macmillan, 1965

P. HALMOS
Introduction to Hilbert Space and the Theory of Spectral Multiplicity, Chelsea, 1951

G. HELLWIG
Differential Operators of Mathematical Physics, Addison-Wesley, 1967

E. HILLE AND R. S. PHILLIPS
Functional Analysis and Semi-groups, American Mathematical Society, 1957

J. L. LIONS
Equations Differentielles et Problèmes aux Limites, Springer, 1961

L. H. LOOMIS
An Introduction to Abstract Harmonic Analysis, Van Nostrand, 1953

J. VON NEUMANN
Mathematische Grundlagen der Quantenmechanik, Springer, 1932

I. STAKGOLD
Boundary Value Problems of Mathematical Physics, Macmillan, 1967

M. H. STONE
Linear Transformations in Hilbert Space and their Applications to Analysis, American Mathematical Society, 1932

B. Sz.–Nagy
Spektraldarstellung linearer Transformationen des Hilbertschen Raumes, Springer, 1942

A. Taylor
Introduction to Functional Analysis, Wiley, 1958

K. Yosida
Functional Analysis, Academic Press, 1965

K. Yosida
Lectures on Differential and Integral Equations, Interscience, 1960

CITED REFERENCES

S. Hartman and J. Mikusinski
The Theory of Lebesgue Measure and Integration, Pergamon Press, 1961

F. Riesz and B. Sz.–Nagy
Functional Analysis, Ungar, 1956

H. L. Royden
Real Analysis (2nd Ed.), Macmillan, 1968

W. Rudin
Real and Complex Analysis, McGraw-Hill, 1966

G. F. Simmons
Introduction to Topology and Modern Analysis, McGraw-Hill, 1963

E. C. Titchmarsh
The Theory of Functions (2nd Ed.), Oxford University Press, 1939

INDEX